Additional praise for *Unraveled*:

"The globalization of the fashion industry has fueled poverty reduction and economic growth in many parts of the world, but it has also left a trail of human suffering and environmental damage in its wake. This fascinating book holds a magnifying glass to the global division of labor to tell us about the 'full life history' of the clothes we wear."
—Dani Rodrik, Ford Foundation Professor of International Political Economy at the John F. Kennedy School of Government at Harvard University, author of *The Globalization Paradox*

"You need to read this book. The fashion industry has become one of the great humanitarian crises of our time. Bédat pulls back the curtain on the industry with devastating insight, simultaneously offering a way forward to a future of industry without gross overconsumption and oppression."
—John Mark Comer, pastor at Bridgetown Church, author of *The Ruthless Elimination of Hurry*

"Captivating. With vibrant storytelling, *Unraveled* weaves together the pieces of a complex system that affects us all but that we can't see ourselves. A revolutionary read that captures both the problems and the solutions needed for a more equitable world."
—Amber Valletta, supermodel and activist

Unraveled

The Life and Death
of a Garment

Maxine Bédat

PORTFOLIO / PENGUIN

Portfolio / Penguin
An imprint of Penguin Random House LLC
penguinrandomhouse.com

Most Portfolio books are available at a discount when purchased in quantity for sales promotions or corporate use. Special editions, which include personalized covers, excerpts, and corporate imprints, can be created when purchased in large quantities. For more information, please call (212) 572-2232 or e-mail specialmarkets@penguinrandomhouse.com. Your local bookstore can also assist with discounted bulk purchases using the Penguin Random House corporate Business-to-Business program. For assistance in locating a participating retailer, e-mail B2B@penguinrandomhouse.com.

Owing to limitations of space, illustration credits may be found on page 297.

Images by author unless noted in the credits.

Library of Congress Cataloging-in-Publication Data

Names: Bédat, Maxine, author.
Title: Unraveled : the life and death of a garment / Maxine Bédat.
Description: New York : Portfolio, 2021. | Includes bibliographical references and index.
Identifiers: LCCN 2020048296 (print) | LCCN 2020048297 (ebook) | ISBN 9780593085974 (hardcover) | ISBN 9780593085981 (ebook)
Subjects: LCSH: Clothing trade. | Labor and globalization. | Business logistics.
Classification: LCC HD9940.A2 .B433 2021 (print) | LCC HD9940.A2 (ebook) | DDC 338.4/7687—dc23
LC record available at https://lccn.loc.gov/2020048296
LC ebook record available at https://lccn.loc.gov/2020048297

Printed in the United States of America
1st Printing

Book design by Cassandra Garruzzo

To my father, Keith, and Jennifer's father, Kevin,
who helped us believe in our own voices

And my daughter, Leontine, and the generations that follow

Contents

Introduction ix

ONE
Growth Mentality: Cotton Farming in Texas 1

TWO
Textiles Made in China: How the Drive for Cheap Is Killing the Planet 22

THREE
My Factory Is a Cage: Cut and Sew and the Crisis of Labor 55

FOUR
Middlemen, Management, Marketing, and a New Kind of Transparency 94

FIVE
Reclaiming Essentials for All: Packing and Distribution 118

SIX
More Is More: Consumerism Goes Viral 148

SEVEN
Tidying Up: What Happens to Clothes When We Get Rid of Them 179

EIGHT
Paved with Good Intentions: The End of the Road for Clothing in Ghana 201

NINE
Let the Makeover Begin: Time for a New, New Deal 231

Acknowledgments 255

Notes 259

Art Credits 297

Index 299

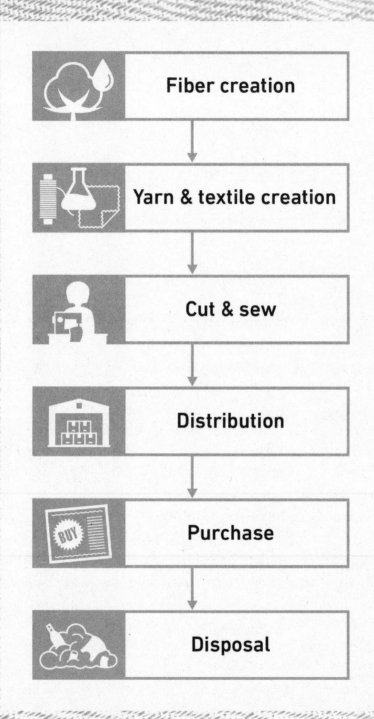

Introduction

illed as "the Marketplace of the World," New York's Jacob Javits Center stood before me like a glass fortress. As a novice in the fashion world, I was intimidated by the prospect of entering the building for the 2013 MRKET menswear trade show, where buyers from across the country were descending to find the goods that would land on their shelves. I grew up in Minnesota, home of the Mall of America, the largest mall in the United States, so I thought I could handle retail on a scale as grand as this. But that morning I felt humbled by what seemed to be the aggregate of all the malls in the world, complete with the familiar convention center aroma of antifreeze and undertones of coffee and pizza grease.

My business partner and I were getting ready to launch our new ecommerce site for clothing and select homewares, Zady. We wanted to present about fifty pieces—shirts, pants, and accessories for women and men—curated from brands that aligned with our philosophy and aesthetic. The goal was to unearth beautiful, artisanal products, so that day in September I was on a mission to find possible goods and brands for the site.

Approaching the enormous main atrium, I felt a chill set in, literal and metaphorical. The other buyers I could see were out for the kill, silently judging and scanning the racks of clothing and signage that stretched as far as the eye could see. Not quite knowing where to start, I walked the length of the space, stopping anytime a collection caught my eye. I approached the first sales representative with what I thought was a simple question: "Can you tell me where your collection is made?" He responded with a blank stare followed by a shrug of the shoulders, then averted his eyes. I couldn't believe it. Why wasn't he answering me? Didn't he know where his own goods were manufactured?

I continued down the aisle until another collection caught my eye. I asked the same question of the woman at the booth. "Asia," she responded shortly, seemingly upset. What was with these people? Asia has forty-eight countries and 4.7 billion inhabitants, so while ever so slightly more specific than the first man's grunt, this information was not terribly helpful. I walked on. It didn't get better: Clothes, it turned out, were being made "abroad," in "the Orient," or to be super precise, "China." What the hell? How did the people responsible for selling clothing to the entire US market not know something as basic as where that clothing was made? (Also, the 1960s called and wants its offensive language back.) Even saying something was "designed in New York," I soon realized, was often code for "made in China." After a long day trudging through Javits with zero business cards and not optimistic for Zady's future, I left frustrated, confused, and skeptical.

Having recently graduated from law school, I was not afraid of research, and quickly applied those skills to find companies that knew who made their clothes. In September 2013, we launched the business with diverse partners, including Imogene + Willie, a company cutting and sewing denim in Tennessee, and Clare Vivier, whose colorful bags were produced in Los Angeles. We interviewed the designers and those responsible for production to tell the story behind every piece we carried, an attempt to give the customer the novel experience of knowing who made their things, down to the individual. We embedded a map on our site to show where the products came from, educating customers more deeply than the brand reps at Javits.

We had thought it would be enough to explain where each article was made—as in, what the label inside the garment would say. But we soon realized that wasn't enough. Our cashmere sweater was "Made in Italy," in that it was where the yarn was knit into a sweater. But the yarn was not spun in Italy and the goats whose wool fiber became yarn roamed the steppes of Mongolia (the cold weather of the steppes helps produce soft fibers). Some of the brands knew where they purchased their yarn or finished textiles, but that didn't mean they knew where those companies, in turn, had purchased the leather or cotton, wool or polyester fibers.

We tried to find a company—one single company—that we could promote through Zady that was addressing these issues of transparency and knowledge, who knew the story of their clothes. But we could not. Not a single company knew the full life story of their garments: from the farm or oil rig that created the fibers, to the factories and people responsible for spinning, weaving, dyeing, cutting, and sewing.

As we dug deeper, we also realized that the tags ignored something else: the environmental impact. Apparel production was increasing precipitously, doubling in the fifteen years between 2000 and 2015; 100 to 150 billion new pieces of clothing produced annually, according to a range of reports. We soon realized that the earlier stages of the production cycle—raw material and fabric production—had the highest environmental footprint, but also garnered the least attention. Fashion companies were just beginning to market products as "sustainable," but it was clear that that word was all but meaningless.

So we ventured to do it ourselves. We made a wool sweater (named .01 The Sweater), with the intention of being transparent and sustainable with every facet of how it would be made. The wool came from a ranch in Oregon, was washed in South Carolina, dyed and spun into yarn in Pennsylvania, and knit into a sweater in California. We launched the sweater alongside a very simple explanation of the social and environmental impacts that all clothing has. The response was overwhelming—not just from customers who were delighted to connect to their clothes, but from other companies, companies much, much larger than ours. Thank you, they said, for putting together the information on the impacts of clothing; it was really helpful for me and my team.

I should not have been shocked, but I was. It wasn't just that customers were ignorant of the impact of the fashion industry. So was the industry, which had blinded itself to its own inner workings—and consequently, its own catastrophic consequences.

＝＝

I got into fashion as an outsider, and still see myself as one. My path into the world of clothing began not in fashion school, but in law school and through work at the United Nations.

One summer, I was dispatched to Arusha, Tanzania, working at the International Criminal Tribunal for Rwanda. I spent my weekends doing my fair share of wandering, and in Arusha on a weekend any wandering will lead you to a market. These gatherings of makers, growers, and entrepreneurs were vibrant, joyful, and irresistible to my twenty-something self. At first, I used these shopping trips to find souvenirs to bring back home.

The more time I spent there, though, the more I realized I was getting more than just gifts I thought my relatives and friends would find pretty. I started to form relationships with the people who were making some of the things that I was buying—the woman who'd made my cool, yet classic floral shorts, the man who wove the baskets I hung on my walls. Visiting those markets, where the fabrics, colors, and patterns were rich with cultural significance, I also came to value the beauty of handcrafted work. Beyond the odd trip to the farmers' market, I really had never had meaningful interactions with the people and places behind what I consumed.

The conversations and discoveries I had at that market would return to me in the months and years after. I started to connect the dots between the Sustainable Development Goals the UN folks were so focused on and the things we purchased. That is to say, environmental degradation and poverty, for example, were both tied to how our things are made and paid for and how we use them. How global trade relationships are structured are a major determinant of whether people have the opportunity to make things and earn a living to begin with. At the same time, the craftspeople I met showed exceptional creative talent and an entrepreneurial drive—they had what it took to make livelihoods that were sustainable for their families, communities, and the planet. What if I could make a business out of working with people like that—artisans whose talents customers would be pleased to know about and support with their purchase power? Eventually, that impulse led me to cofound Zady—which then led me to the Javits Center, confronted with the realities of the fashion industry.

As Zady evolved, I kept digging deeper. If I was really most interested in advancing "sustainability," was the best vehicle to do that a company whose business model, when it came down to it, was to sell more things? And since citizens and the industry itself had proven to not know its own impact, could I play a more helpful role by bringing that information to light? Eventually, I decided to stop selling clothing and focus all my attention instead on explaining the fashion industry's real impact. I connected with experts who understood every aspect of fashion's impact on our world—agronomists, climate scientists, historians, fashion executives, factory executives, and material scientists; labor experts, organizers, and laborers; political scientists, toxicologists, psychologists, marketers, and economists—to launch a research and action think tank—a step above your average think tank, where ideation stops at the whiteboard. The New Standard Institute (NSI) is an effort to use information, data, and stories not for private profit, but for the public good.

My goal at New Standard Institute is to provide rigorous research and data (and highlight when more is needed) about the fashion industry, which is not, unsurprisingly, known for transparency. As we will cover in the pages ahead, the processes and practices behind how our clothes get made have flown under the radar. As a result, the information that has been captured thus far has been piecemeal and often inaccurate.

Data is one way to tell a story, and a pretty convincing one at that, but most of what we all do has little to do with data. (If we all acted on data alone, we'd be in a very different world right now.) The stories that get us to act are ones that spark something in our spirits, that put up a mirror to our own experience that reminds us of others who have the same values, fears, triumphs, and dreams as ours. What you are about to read is that kind of story.

═══

When I set out to write *Unraveled*, I wanted to trace the life and death of a single pair of jeans—a garment that is ubiquitous in our culture, beloved for function and style alike—from farm to landfill. This is an extension of the journey I began at Zady, trying to tell the basic origin story of a garment. But there were obstacles, as I had already discovered with Zady. Companies do not have a clear understanding of their own supply chains, and many manufacturers are not exactly willing to throw their doors open for scrutiny. These roadblocks show just how far the industry has to go before it achieves anything close to true transparency. So while this story does not follow one single pair of jeans in a literal way, it does follow where an average pair might go, alongside many other kinds of garments (everything goes with jeans, right?).

In the story that follows we will visit cotton farms in Texas, which was and still is a significant source of global cotton production, meeting farmers navigating the trade-offs between the health of their land, their bank accounts, and themselves. In China we will see how those raw fibers are spun into yarn, dyed, and woven into denim. And in Sri Lanka and Bangladesh we will meet the women responsible for cutting and sewing fabric into a final garment. Back in America, we will go inside an Amazon warehouse to see how our jeans are shipped and make their way to our closets. Finally we travel to Ghana, where quite a bit of our clothing lands after we've had our way with it, becoming our jeans' final resting place.

The story of a pair of jeans is the story of modern fashion and capitalism, another reason why they make a particularly fitting hero for our journey. Today, 1.25 billion (yes, that's billion with a "b") pairs of jeans are sold globally every year, and the average American woman has seven pairs in her closet. They are evidently a big player in the fashion world, which is itself a major player in the global economy. The jeans we wear today have become an ironic symbol of democracy that even American presidents have played into (until the suit-donning forty-fifth, that is). They are billed as all-American, but the truth of their creation takes us far outside the US borders, and deeper inside where we'd ever think to look. The story of our pair of jeans will take us around the world and back, reflecting the sprawl of our supply chains and the degree of cultural fusion that has allowed fashion to become the radically opaque and exploitative force it is today.

I could give you all this in numbers and charts, and there will be a few of those in the pages ahead, but more important this book introduces you to the people who are involved in making your clothes. What their stories reveal is that understanding the systems of creation and distribution of clothing, of how they are marketed and the impact that marketing has on us, helps us understand our broader world and our role in it.

Until very recently, this $2.5 trillion industry has been relegated to the "style" section, connoting that it's superficial, girly, fun, unimportant. Yet it's an enormous industry. It is responsible for the incredible net worth of a few people at the very top of the list of the wealthiest people on the planet. It employs millions of the most vulnerable people globally—a majority of them women—and engages some of the lowest-paid labor domestically, as well. And it has a significantly destructive impact on our environment, contributing, according to one report, at least the same level of greenhouse gases as France, Germany, and the United Kingdom combined. The roots of how we dress are also the roots of slavery and colonialism—systems of oppression that we will see are far from fully dismantled, and behind the conflicts over racial equality that are raging today. Our deeply unequal economic system is also the fruit of those systems. Seen together, the story of our clothes also helps us understand why our societies have become so divided. In the words of historian Sven Beckert, "Too often we prefer to erase the realities of slavery, expropriation, and colonialism from the history of capitalism, craving a nobler, cleaner capitalism. We tend to recall industrial capitalism as male-dominated, whereas women's labor largely created the empire of cotton." Part of my intention in writing this book is to put fashion and the clothing industry in its rightful place as not just part of, but at the foundation of, industry and society as we know it.

I have come to believe that the reason this industry has not been taken seriously in the world of policy and business is because it has been relegated to the domain of social "minorities"—namely, women and people of color (and often both). Since the earliest days of industrialization, clothing has largely been made by people belonging to both these two groups, and marketed mostly to women; we'll meet some of their progeny—the people whose hands make our garments today—in the pages that follow. Even in environmental circles, fashion is very often dismissed. I can't tell you the number of times I have spoken with significant environmental donors about the impact of the fashion industry during my work for NSI, to which they would respond, *Oh, you should talk to my wife about that, she loves fashion.* (That would of course be meaningful if the wives of major environmental donors held the purse strings, but from my experience, they do not.)

The lack of attention has allowed the industry to operate with very little regulation and not as significant coverage in the media, all the while its (mostly male) executives make enormous sums of money on the backs of (mostly) women's work and women's purchasing. I write this book as someone who both is deeply troubled by the industry and the society that has been created in the effort to sell us more of it and appreciates the power and pleasure of clothing. The knowledge of how our clothing is made, marketed, sold, worn, and discarded is a powerful lens through which to better see the truths of our world and its history, however beautiful or ugly they may be. Seeing clearly is the first step to dismantling the urgent injustices this book describes and bringing about not only a more just society, but a pleasurable and thriving one.

Denim has a global history, intertwined with the rise of modern capitalism. That story begins in India, where cotton has been grown and worn as clothes since around 6000 BC. In the seventeenth century, on the shores of what is now Mumbai, impoverished workers in the port city of Dongri donned garments made of a thick, coarse, cotton fabric called "dungri." Cotton in any form was completely foreign to Europeans and Americans, and when they saw it they went bananas. Think your high-waisted jeans and cheap cashmere sweater are uncomfortable? Try living your life in a rotating wardrobe of animal skins, wool, and linen, which is pretty much all Westerners knew until the 1800s. It was not a comfortable, or particularly colorful (those materials don't hold dyes so well) time.

Awestruck by cotton's unprecedented softness, lightness, and durability—*the touch! the feel!*—Europeans at first couldn't decide if it was an animal or a plant (they called it a "vegetable lamb"). But what they did know was how valuable this white fluff was. Intoxicating an entire continent, "white gold," as it would come to be known, led to colonialism, the expansion of slavery, the rise of Europe, and the creation of the capitalist system and institutions from which we still operate today. Lusting for clothes that smelled less and gave them fewer rashes, all of the major European powers—the Dutch, Danish, French, and British—got in on the lucrative cotton trade, which was still rooted in India, by each forming their own East India Company. In order to ensure a steady stream of the stuff that would become the fabric of our lives, the Europeans began spinning a network of trade routes that crisscrossed not just India but Africa, Europe, and eventually the Americas.

This required enormous amounts of capital and the institutions surrounding that capital—banks, contracts, lawyers, corporations, and government institutions that could ensure contracts would be enforced. In the years before the United States declared independence, cotton textiles accounted for three quarters of the East India Company's exports. This giant of a commodity would fundamentally change how people interacted at every level of society. Without cotton cloth, we would have no global economy, no staggering social inequality between the Global North and South, no work for women outside the home, and no industrialization, which was all powered by slavery on expropriated and overtaxed land.

Indeed, during the preindustrial period, cotton may have been a major innovation in terms of comfort and washability but remained wildly labor intensive. First there's the labor of picking, performed in America mostly by African people who were enslaved and brought to the southern states. But the work doesn't stop there. Cotton fibers also need to be removed from tightly wound and prickly seed pods (called bolls) in order to be transformed into jeans, or anything like them. By hand, it could take one person ten hours to separate one pound of fiber from the seeds. But then, or as we were taught, along came Eli Whitney, who thought he could make the process more efficient. He patented the cotton gin in 1793, a machine that mechanically separated the seeds from the boll, allowing someone to process about fifty pounds of cotton a day. While it was Eli Whitney that is given credit for inventing this process, historians now believe that the initial ideas came from uncredited African people who were enslaved, who, not being allowed citizenship, were not legally allowed to hold patent rights. With the creation of this production system, the Industrial Revolution commenced, and the world as we (rather, they) knew it had changed forever.

The newly industrialized gin process only increased the already insatiable demand for cotton. Cotton *picking* had not yet been industrialized, however, so in order to get more cotton there needed to be more people in the fields. As a consequence, the institution of slavery expanded. Cotton yields in the United States went from 1.5 million pounds in 1790 to 2.275 billion pounds by the eve of the Civil War in 1860. Southern agricultural regions thrived off this industry. Southern port cities like New Orleans and Charleston would not be the major cities they are today without the ships full of white gold they sent to Europe.

Cotton wouldn't have been such a successful coup for its owners, financiers, and customers without slavery. In 1850, 3.2 million people who were enslaved worked the fields of fifteen slave states, 1.8 million in cotton fields specifically, and their lives are woven into our national and social fabric. The South was not alone in its support of slavery. While outlawed in the North, slavery was financed by Northern (and British) merchants, bankers, and investors, and their cooperation in order to satisfy conditions of supply and demand is the root of modern-day capitalism.

Things did not have to be this way. Consider how all of the other major cotton-growing countries of the time (and today), such as India, did not turn cotton into white gold using the depraved means the United States did. American cotton plantations also established the first forms of formal workplace management, with unsettling consequences for modern workers. Slaveholders kept detailed track of workers' output, rendering humans into cotton-picking machines and valuing them based on performance—a theme that you will see is now embedded throughout the production and distribution of our clothing today.

Going back to our process, we find even more exploited work and commodification taking place after the gin. Once cleaned, those fibers needed to be transported to the cotton-hungry people of Europe, as well as spun into yarn and woven into fabric before we could make them into jeans, or dungarees, or whatever else you want to call your pants. This, too, was arduous labor, which in preindustrial times had been done by women in the home, who sat at spinning wheels turning wool into yarn for their families' needs. Post–cotton gin, though, the demand for spinners and weavers increased to meet higher demand for textiles, creating yet another stream of labor capital.

Cotton drove modern industrialization and inequality in this way. The higher those with access to cotton rose on the socioeconomic ladder, the lower those who didn't fell. The former was now in a position to derive labor from workers in a system that relied on inequality—there was always a taker and a giver, a winner and a loser, despite constitutional claims (and balanced books) to the contrary. Thanks to cotton, humans became commodities the world over; Indian cloth was used as currency with African traders in exchange for people who were enslaved, who in turn were used to expand production in the American South and make more cotton (and cloth) to sell somewhere else.

To maximize the potential of the gin, plantation owners needed more quality cotton-growing land, which they found farther south and out west. The only thing standing in their way was, well, people—this time, the indigenous people of the Americas. The government ensured that it would stay in the cotton race by forcibly displacing people from their land in a trail of death and destruction. In fact, the Louisiana Purchase, which created many of these cotton-growing states from Native lands, was structured by none other than Thomas Baring of Britain, one of the world's leading cotton merchants.

As we will see in the chapters that follow, similar trade flows are still intact today. Legal slavery has been abolished, but the exploitation of labor and land still continues. One of the promises of capitalism—which, as we know, fashion helped create—is shared prosperity. Yet according to the Edelman Trust Barometer in January 2020, 56 percent of respondents found that the economy in its current form was not working for them. By 2020, when I completed the majority of the work on this book, a tsunami of conflict, upheaval, and loss had made the notion of "shared prosperity" seem utterly fantastical. The pandemic and its related societal and economic quakes exposed the fragile seams of our global fabric woven from exploitation and deception. And it laid bare how manufactured our desire and need for more is, and how quickly our craze for clothing can shift once the stakes get higher. At one point, jeans may have represented an ideal of democracy and equality, but the jeans our society is wearing have become frayed to the point of distaste. If we want to reclaim true democratic values, we need to reexamine how our political and economic systems are woven into the clothes we buy, wear, and discard.

We are living in extremely challenging times—culturally, economically, environmentally. In this moment of crisis, it has never been more important to understand how the actions we take can impact the world—for worse or for better—and what we can do to turn our clothing, ourselves, and our institutions into a force for good. In the last chapter of this book, I offer more specifics on how clothing can be a gateway to reclaiming our citizenship, but before we get there we need to meet a few people. So let's go.

Unraveled

One

Growth Mentality:
Cotton Farming in Texas

When you think of starting a story about clothing, you might imagine a glossy urban center, people meandering through a skillfully designed maze of racks and mannequins and signage, or a minimalist website with an infinite scroll of polished product photos. But behind all that product and marketing, the origin of our jeans, whether it's a cotton farm or an oil rig (but more on that later), looks nothing like the place where you buy them. And as I stood before the rolling acreage of Carl Pepper's farm outside of Lubbock, in the northwest of Texas, on a warm October morning, the contrast between the luxury marketing and the realities of nature—weather and luck are the difference between success or failure—could not have been more stark. There's a whole world, full of people, plants, microorganisms, chemicals, and carbon, that also gets bought and sold each time we say "yes" to a new item of clothing.

When it comes to cotton production, the United States comes in third worldwide, after China and India, thanks to the pillowy fields of white fluff like those on Carl's farm. More than half of all land used for cotton in America is in Texas, and that land produces 40 to 50 percent of all the domestic harvest—in 2019, 8.8 billion pounds are estimated to be grown stateside. Worldwide, it's grown in eighty different countries, and 2.53 percent of all land used for agriculture is planted for the sake of our jeans (and other cotton things). In 2019, the USDA estimated worldwide cotton production reached 58.5 billion pounds.

Carl, my guide for all things cotton, is no gentleman farmer, but a farmer who is a true gentleman. Of average build, with dusty blue eyes and sporting a baseball cap and long sleeves to protect from the harsh Texas sun, he gave off a good Christian farmer vibe but with a renegade spirit and a keen mind for numbers. His words were easy on the ears—they were delivered with a light southern drawl that I found so appealing that I actually had to stop myself from trying to mimic it.

Carl Pepper

The Pepper family has been in the United States for centuries. Carl's father's side landed in Texas in 1876 by way of Virginia, Kentucky, and Missouri, with some family having arrived in America as far back as the 1700s from what is now Germany. He was born and raised in South Texas, to the west of San Antonio, on an 800-acre irrigated, chemically intense farm. He told me that his father, Leslie Pepper, was among the first generation to use these industrial farming practices to try to enhance crop yields. Carl's father embraced the latest and greatest science to try to tame nature. New chemical products and irrigation tactics helped grow his farm into a successful business that supported his family. Leslie's success even persuaded his sons to follow in the family business. Carl recalled how his father thought nothing about spraying his fields with the new artificial fertilizer with a boom in front of the tractor, completely unprotected from the heavy mist of chemicals—it's just what they did. Chemical farming became popular starting in the 1940s, when innovations and supplies for the war were translated for agricultural use. Traditional farming practices, which included the rotation of crops and letting land be fallow to allow for nutrition to be built back into the soil, became displaced by extensive use of chemical fertilizers and pesticides.

It was these chemical practices that Carl inherited when he first took over his father's farm twenty-seven years ago. His inheritance came a bit earlier than expected, though, when, after around thirty years of chemical spraying, in 1987, Carl's father was diagnosed with acute leukemia. He died three days later. On his deathbed, he gathered Carl and his brothers and told them, "Boys, I'm not sure what I did to myself with all these chemicals."

As we toured his field in his white Chevy Silverado, Carl recounted his father's last words with surprising calm. He had clearly reflected on them—and his father's life—often. "That started a thinking process," he continued.

In 1992, Carl had an opportunity to relocate from South Texas to Lubbock. His middle brother, Terry, had come into a piece of dry land—agricultural land that does not require irrigation, and only uses rainwater—that his wife had inherited from her grandfather. Terry's wife was environmentally conscious and was eager to convert the land to be certified organic, a standard that the government was rolling out that year, thanks to much citizen demand.

Carl wasn't entirely sold on the whole organic thing. West Texan soil is naturally arid, so there was no guarantee that his cotton would grow without chemicals. But with his dad's last words echoing in his head, he thought it worth a try. If the organic route didn't work out, there was no shortage of chemicals that promised to revive unproductive fields like a Chia Pet. The team started with 160 acres along the highway. At the same time, Carl's older brother, Kelly, founded the Texas Organic Cotton Marketing Cooperative, of which he is still the manager. Comprising thirty-five farm families, up until three years ago the Cooperative was responsible for 2 to 3 percent of organic cotton production in the world, and a full 85 to 90 percent of United States' organic cotton yields.

Cotton grown with a heavy dose of chemical assistance, like Carl's dad and he himself grew, is now, after only a generation, known as "conventional" cotton—a misnomer if there ever was one. I was surprised to learn that the widespread acceptance of "conventional" was such a recent development, a fact made more ironic as I looked out over the neat rows of Carl's cotton bushes growing freely and abundantly without the crutch of synthetic fertilizer and pesticides. We now give a special name to this unadulterated form of farming—organic—but nature's way is the more conventional way of doing things.

GOING GREEN: THE FINANCIAL COST OF ORGANIC COTTON

When you think "organic," what comes to mind? Perhaps "better for me," "better for the earth," "sustainable," or maybe even "green." Given all of these seemingly great properties, why, I wondered, were more farmers like Carl not converting their fields? It turns out, some of that may be true some of the time, but as we know from Kermit the Frog—and, it turns out, Carl the farmer—green, both environmental and financial, is not exactly the easiest thing to be. Nature may be the OG organic, but the garment industry today doesn't care all that much. As Carl explained over the course of our time together, when you farm organic, especially in the beginning, the farmer must bear additional costs. When you buy organic milk or organic cotton produced in the United States, that label is regulated by the government through the US Department of Agriculture (USDA). A product produced under the USDA organic standard is generally made without synthetic chemicals (though there are a few exceptions) and with non–genetically modified seeds. Keep in mind that a product made with organic cotton has no bearing on the chemicals that are used after the farm, a part of the journey we'll see later.

It takes a minimum of three years for a farm to become certified organic and costs around $1,500 for the certification fee. The USDA requires three years' worth of land-use records demonstrating that no restricted herbicides or insecticides were applied in that time as proof that the chemicals are sufficiently removed from the soil. Once three years and a day have passed, anything harvested off that land can be certified organic. The process to convert the land requires an openness to new methods and practices, such as crop rotation to naturally revive the soil, and more staff to manually pick the weeds that conventional farmers just douse with herbicide. So while expenses go up during the three years of transition, the farmer is not able to sell his (or her, but most likely his) product at a premium to help recoup his investment. Carl's original lot was already relatively low in chemical use when he sought certification, so he had a relatively gentle learning curve in changing practices and acclimating himself to fickle, untreated crops. But still, nature's schedule for "cleansing" itself of chemicals wasn't up to speed with the demands of a small business owner like Carl.

Even if a cotton farmer is willing to take on the effort of learning new techniques and the financial risk, the payout may or may not be worth it. The market price for harvested organic cotton has fluctuated downward relative to conventional cotton. Carl explained that most organic products—namely, food—bring in double the price as nonorganic. In 1992, when he started, organic cotton followed that same trend and was $1.00 per pound (conventional was going at about $.58). "Yee-haw, I thought I was rich," he said with a toothy grin, his blue eyes creasing at the corners. A drought in the nineties lowered prices to around 80 cents, but prices shot up to $1.50 into the early 2000s, which allowed Carl to make a good margin and cover the cost of equipment upgrades at the barn.

But times are changing. Organic prices are being squeezed down to around $1.00 to $1.10 once again because of growing competition from some other domestic farmers who converted and a huge influx of organic cotton from India and Turkey. Meanwhile, even though organic is itself a teeny tiny portion of all cotton produced (just .7 percent), there does not seem to be enough demand from the clothing or cosmetic and personal care industries for even that very small supply.

I was beginning to see how becoming organic was not such an obvious thing to do from a farmer's perspective. And then Carl shared with me how much inherent risk there is in farming in general. A variable market isn't the only reason organic can mess with a farmer's income—nature herself is just as unreliable. I visited Carl in October 2018, in the midst of a drought. Typically the wet season of May through September would bring about eighteen inches of rain, but 2018 fell about four inches short. Carl's average yield is five hundred pounds per acre, but that year he would only harvest fifty pounds an acre, leaving him shy of his projected income for the year by tenfold.

Farmers have been battling unpredictable yields, and unpredictable profits, since agriculture began. It's part of being a farmer, Carl said, and one of the reasons it's understandable so many drink and take drugs to relieve stress. The suicide rate for male farmers in seventeen states was double the rate of the general population's. Given all the inherent risk, I could begin to see why farmers would not immediately jump at the opportunity to convert their fields. Still, Carl's attitude is to make lemonade out of lemons; boom years are when he can keep up with other projects, like repairing and upgrading equipment, but in dry years you "just sit tight, hold through, and make it work." During these periods he's humbled enough to remember to "stop and look around and say, 'Who's in charge? I'm not.'"

The cotton boll is the fruit of the cotton plant, and this is the cotton flower.

To battle the variations in price and output, the cotton industry has successfully lobbied for crop subsidies, which has amounted to an average government transfer to cotton farmers of $2.1 billion annually since 1995. The subsidies are primarily twofold: First is a guaranteed floor price for cotton, which takes the form of a payment the government gives to farmers when cotton falls below that price in the global market; second is government-financed insurance to protect against the vagaries of nature. Subsidies are a good reminder of how, when people (like cotton farmers) rally their political power together, they can get the government to institute policies in their favor. And these subsidies are a significant reason that cotton is even grown here; there are a lot of other places in the world where it's cheaper to get our cotton.

Advancing technology has made the farming process, organic or

conventional, faster and, I admit it, kind of cool. I timed my visit to
Carl's farm with the harvest, so I was able to see it in action. In South
Texas, farmers can harvest as early as July. But in West Texas, the harvest begins in late September or early October. Enormous machine strip-
pers that look like giant vacuum cleaners pick the cotton boll from the
plant. This was once done by hands that were enslaved in the American
South; now, it's still done by hand in countries with dangerously low
labor costs, or by forced labor, including in Xinjiang, China, as recent
reports indicate. As the cotton is stripped from the plant it is collected
into a large cage that makes up the bulk of the stripping vehicle. Once
full, the stripper dumps its load into another machine vehicle that com-
presses the cotton coming in from the stripper into a loaf, so named
because of its resemblance to a giant loaf of bread. Each loaf can weigh
up to 25,000 pounds. With these machines, Carl can harvest around
four thousand acres with just four workers in three or four days' time.

Stripping cotton in West Texas.

The cotton loaf, arriving at the gin.

Farmers are then responsible for hauling the loaf by semitrailer to the neighboring gin, where the fibers are separated from the seed and both parts are sold. The seed is used for animal feed, or it's pressed for cottonseed oil to be used in cosmetics or cooking. You know the Crisco cooking brand? Crisco stands for "crystallized cottonseed oil." After ginning, the cotton fiber is stored in a warehouse at the gin until it is sold to a mill, where the fibers will be spun into yarn that can be woven into fabric like denim or knit into a T-shirt. While there used to be a mill in Littlefield under five minutes away from the gin, it was sold for $250 million to a milk processor in 2015. Most of Carl's cotton now goes to Mexico and Asia, first via railcar and then through the ports in California.

For most conventional farmers, the buck stops at the gin. In Lubbock, the gin is the proverbial watering hole for local farmers, one of the places to socialize other than church, but after the cotton loaf is dropped off it's out of sight, out of mind—and one of the ways any sort of "transparency" in the system is clouded quite early on in the process of making a garment. That's because many farmers are part of a cotton exchange, which splinters off cotton's journey into many more, often unnecessary, hands. Farmers in exchanges have no idea what happens to their cotton after it goes to the gin—it might end up as a pair of jeans or a baby sock, it might get mixed in with fibers of a very different category or an organic/conventional blend. The gin, warehouse, and marketers who triangulate the transactions might all be different entities, and when each of them takes a cut it lowers farmers' earnings, increases the price of cotton products, and stops information sharing between cotton growers and cotton buyers.

Here we have a system that attempts to simplify, even improve, the business of farming and selling cotton, but by design has backfired. With more steps instituted to ensure that the farmer only has to care about growing as much cotton as possible, we've weakened and complicated the entire system, and guess who's left to foot the bill? The farmers who manage the land.

The Cooperative has proven to be Carl's saving grace. It was originally conceived as a way to pool production to stabilize prices and market to manufacturers, which Carl's brother is in charge of, for the long- and short-term benefit of organic farmers. This happens in two ways. First, after ginned cotton is sent to the warehouse, the Coop handles both marketing and delivery to clients; without outside brokers, costs immediately go down. The system also helps to maintain better communication between growers and customers. Some of these clients go back twenty years or more, and while they're mostly niche players, their long-standing relationships are valuable. The farmers all know their customers personally and are able to share predictions about the yield and quality of a given harvest, which makes planning easier for their customers. Smaller orders from these clients make for smaller invoices and the need for more customers, but it is also a defense mechanism for the farmers long-term; as any good financial adviser would say, diversity is the key to stability.

The Cooperative's diversifying strategy proved successful when, a few years ago, the board members took a meeting with a major company (Carl asked that I not name it), whose executives had made a grand entrance arriving in their private jets. "We'll take everything you've got," they said, painting a picture of an enormous boon for the Coop that anyone would be hard-pressed to ignore. But, before signing on, they found out about a deal the company had made with some nearby cucumber farmers to make pickles. Wooing the farmers with a similar big cash buyout, the company wound up selling gallon jars for the price of a quart and passed the cost of those savings to the farmers. Needless to say, this soured the deal for the farmers. Talking a big talk, the company also walked a big walk away when pickles didn't work out, leaving the farmers with virtually nothing. With this in mind, the Cooperative decided not to sell to that company, despite their pizzazz. And the Coop members decided to never promise more than 50 percent of their crop to any one client. Once you own over half, you basically own it all.

I checked in with Carl recently to see if anything had changed for his farm during the COVID-19 pandemic and all the economic uncertainty it has wrought. Health-wise, everyone was fine. "We invented social distancing before it became the norm," he said. While he was still dealing with the drought conditions that had plagued the land when I was there, he did not suffer from a lack of demand. Perhaps this is because of the close relationships he has developed through the Cooperative with his vendors, or perhaps it's because most of his vendors are nonclothing companies, which have less variable demand.

If we as consumers want to encourage better cotton practices, we need to actually behave in a way that would tip the scales back in Carl's favor. We need to show brands we want those practices by seeking them out when we do make clothing purchases, and expressing our desires in other ways, too. We can also take the same tactic as the farmers who lobbied for subsidies in the first place and demand that legislation is passed to create financial incentives for farmers to make the switch to more sustainable practices. With political leaders who support more sustainable farming practices, we and the planet we live on might have a fighting chance to harmonize, rather than attempt to but ultimately fail at dominating nature.

GOING GREEN(ER): THE ENVIRONMENTAL COST OF ORGANIC COTTON

Why, despite all of the challenges, have Carl and sixty-seven other farmers in the United States, plus 219,000 farmers worldwide, decided to grow organic cotton? The market for this synthetic-chemical-free version of the fiber that has clothed us, and shaped our economy, for centuries is, despite the challenges, undoubtedly on the rise. The Organic Trade Association's 2019 Organic Industry Survey reported that organic fiber sales increased by 12 percent from 2017 to 2018, reaching $1.8 billion, with most of those sales in organic cotton. As we've seen, though, the organic standard doesn't necessarily provide the answers farmers are looking for from a financial perspective. Organic cotton farmers face significant risks, from the drawn-out certification process to the shrinking organic premium, to the unreliability of the harvest. And, for all this work and risk, researchers are finding that the organic label is not a perfect arbiter of sustainability.

To unpack this quandary more thoroughly, we need to understand a bit more about the biology of cotton, and farming in general, and how we've become reliant on synthetic pesticides and fertilizers. To put the process of farming very simply, seeds get planted, crops grow, and farmers harvest them, removing significant amounts of nitrogen from the soil. Nitrogen is critical to farming, as it gives the plants the energy to produce fruits (the cotton boll is the fruit) or vegetables. Traditionally, farmers let the land rest for a bit, which naturally replenished the soil, or they planted other plants (called cover crops, which usually have less economic value, like oat or buckwheat) that helped rebuild the nitrogen into the soil.

And then industrialization happened. First, during World War I, industrial farming practices included mechanical tilling, removing the cover crop, and continuous farming, leaving no time for the land to lie fallow. These practices made the soil dry and prone to erosion, which led to the Dust Bowl that devastated most of the American West. After World War II, increased demand strained the land further. Returning American soldiers (and their baby-booming families) and international demand for food and clothing from war-torn countries like Germany increased the need for cash crops like soybeans, corn, and cotton, which have the ability to feed and clothe large numbers of both people and livestock. Add to the equation increased access to perishable foods thanks to the rise in automobiles, refrigeration technology, and supermarkets, and farms faced such demand that they were pushed to operate at an even higher capacity. They shortened the amount of time that fields lay fallow, then eliminated it entirely. Crop rotation was dropped, too. Like the soldiers, civilians, and prisoners around the world who had starved during wartime, the soil was struggling to maintain its delicate balance of nutrients.

The problem wasn't exactly the same as in the 1920s and '30s, though. War factories appeared to come to the rescue of struggling farmers through a most surprising way: bombs. No actual explosives were involved in the boom of "conventional" agriculture, but their ingredients have been essential in the rise of chemical-intensive growing practices that have set new standards for the quantity of our food and our natural fibers. During the 1930s and '40s, huge factories had harnessed nitrogen (and its by-product, ammonia) to make bombs for the military. After D-day, with war demand all but ceased, those factories stockpiled with nitrogen found a rather ingenious (though perhaps short-sighted) solution: converting all that ammonia-rich nitrogen into synthetic fertilizer.

This development was revolutionary. Farmers got a way to replenish nitrogen in their soil artificially that didn't require keeping the land fallow or using cover crops. Synthetic fertilizers not only perked up tired soil but seemed to make it better than before; chemically treated farms produced more and bigger crops than ever.

Chemical insecticides (to control bugs) and herbicides (to control weeds) were developed around the same time, beginning with the well-known pesticide dichlorodiphenyltrichloroethane, or DDT. Its insecticidal properties were identified in 1939 by Paul Müller, a Swiss chemist who would win the Nobel Prize for his discovery. DDT was yet another answer to the world's demands for more. DDT kept crops pest-free, which increased yields. Off the farm, it was used as a delouser for returning World War II soldiers and as mosquito repellant in suburban backyards. A 1947 ad proclaimed "DDT Is Good for Me!," with an image of a singing milkmaid beside her cow, an apple, a potato, a rooster, and a dog; it ex-

plained that DDT could (and should) be used everywhere, from the farm to the home.

But pests would quickly become resistant to DDT, requiring ever greater quantities to achieve the same results. Not until Rachel Carson's seminal *Silent Spring* was published in 1962 were the detrimental effects of DDT exposed, including liver tumors in animals and potential reproductive effects in humans, making us second-guess soaking our food, soil, and bodies in this toxin year after year, meal after meal. It was banned in the United States by the EPA in 1972. Not exactly the kind of explosion we might have wanted from those bomb factories.

These practices of intensely fertilizing, spraying, rinse, and repeat were what Carl's father learned as "best practices" to make his farm run well, and indeed what Carl was brought up with. Of all the chemicals used for agriculture worldwide, often the most hazardous are doused on fields growing cotton. In one survey of the countries that grow 90 percent of the world's cotton, all listed at least one hazardous pesticide as routinely used in cotton production. Behind just one T-shirt made with conventional cotton, there's one third of a pound of chemicals; there's three quarters of a pound in one pair of jeans.

Some farmers today see little reason to change these ways, and the past several decades have only increased the options farmers have for chemically altered crops, especially GMO seeds like Roundup Ready, which are engineered to be resistant to the potent herbicide Roundup. These farmers include Kent Kahl, who's based in Tahoka, about thirty miles south of Lubbock. Globally over the last eighty years cotton production has tripled thanks to industrial farming—synthetic insecticides and fertilizers but also irrigation and genetically modified seeds. Although he complained about pesticide-resistant cotton seed costing more, Kent noted that using that GMO seed gives him bigger yields (though the evidence on this is not strong in the long run) and that pesticides reduce labor expenses, since he doesn't need to worry about weeding, which is done by hand; compare Carl's four workers managing four thousand acres to a conventional farmer, who can manage three thousand acres with just one man on the job.

Talking with Kent, though, it was clear he was opposed to change in a bigger way, too. Everything about him said "southern gentleman," from the way he carried me over a muddy patch to his refusing to let me pay for dinner at his wife's restaurant. (On the menu? Cow testicles, I am not lying.) He called himself a "redneck" more than once, and was content with doing things the way he's done them all his life, including shipping his cotton off on a boat from Houston to who knows where for processing. That said, he knew of Carl's organic approach—they are friends—and he admired his success in organics. "I don't think Carl's crazy at all," he said. "We spray [insecticides and herbicides], we don't spray as much as we used to. I used to do a lot of it commercially [spraying for other people]. I should have done what he's done years ago but I was scared." When I asked why he was scared, he told me, "I don't know, I guess I was lazy, I'd rather spray."

That's exactly what a lot of farmers in Carl's area thought around 2010 when many farmers were using Roundup Ready cotton, the genetically modified crop that can survive being sprayed with the herbicide Roundup, which was patented by Monsanto (now part of the Bayer company) in 1973 and was the most widely used herbicide in the United States in 2001. They sprayed around 80 million acres of US farmland with glyphosate, the main herbicide found in Roundup. But the land was promptly doused with a heavy rain and wind. There was so much fine particle dust in the air from the Roundup that, thanks to the rain and a southwest wind, a lot of it found its way onto trees and other non-resistant plants far outside of the area originally sprayed. This light, indirect application of herbicides made all the trees sick and "goofy-looking," Carl said, and yet no one at the time was putting the pieces together. "You mess with nature and it will kick your butt," he told me with a laugh. But the butt-kicking we're getting now is no joke.

Chemicals used to grow our cotton are anything but "conventional." And nature, whether it's the land itself or our bodies, does not support the type and quantities of chemicals used. Carl's father was not the only one to succumb to cancer after years of exposure to the stuff; twenty years later, his middle brother Terry was diagnosed with brain cancer. He completed a year of treatment, but was soon physically incapable of walking to his barn. He died two days later, one morning in mid-November. Early that morning Carl woke up suddenly with a strange feeling. His sister-in-law called two hours later with the news that Terry had died, at the same hour Carl had woken up. "Terry's hanging out with Jesus while I'm fighting weeds," he said.

Cancer is a big part of why Carl is an organic farmer today, but the lessons he's learned from his family's ordeal have brought him to a different conclusion about best practices when it comes to synthetic chemicals. On the one hand, not using synthetic chemicals—that is, organic—can help to improve soil health and thereby prevent the need for fertilizers and pesticides to begin with. Within five or six years of being organic, Carl said, you can start seeing changes in the soil—its crumble, ability to retain moisture, nutrient level, even the smell are all different. You don't have to be a farmer, with an eye trained for dirt, to notice these differences. I was shocked by how lush organic farms looked compared to their dustier conventional neighbors. Left to its own devices, and nourished with crop rotation and cover cropping, the soil rebuilds its ecosystem. Microbes that make it disease- and insect-resistant flourish and create nutrients; this is the same bacteria that's in our guts, on our skin, and basically everywhere, and that is responsible for regulating all our bodies' systems include the immune system. The National Soil Health Institute currently has nineteen different metrics to measure soil health, including the presence of organic carbon and other minerals, crop yield, texture, water retention, and micronutrients. It doesn't take much to make a big difference; just 1 percent more organic material increases soil's water retention by twenty thousand gallons per acre, which makes it more resilient in times of drought and flood.

So after all that he has been through, does Carl the organic farmer think the future is organic? Not entirely.

To Carl, soil health and yields are the most important factors for the future of his farm, and he believes it is these metrics—not the stark yes or no, synthetic chemicals or organic—that we should all be turning to for a more sustainable cotton industry at large. Data supports this from another angle, since on average, organic farms yield less than their conventional counterparts. When those yield differences are taken into account, conventional farms, again on average, have a lower carbon footprint per unit of production. I say the word "average" twice in the previous sentences because there are significant differences between individual farms, which suggests that the focus of research should be less on organic versus conventional and more on the specifics of how to get the most yield over time with the lowest carbon footprint, all while not putting farmers, or people who consume farm products (which is all of us), at risk, and on the financial incentives to align farmers with these goals.

Carl advocates for regenerative agriculture, a model that focuses on soil health. He also likes the idea of a "surgical" use of synthetic herbicides, which the organic standard does not allow, since pulling weeds out by hand significantly increases costs for the farmer. With a glimmer of his tinkerer side in his eye, he told me excitedly about a partnership between John Deere and a company called Blue River Technology. They're developing artificial intelligence to give sprayers this capacity, which would cut herbicide use by 90 percent for conventional farmers while saving more diseased crops. But the technology is still very expensive. "I believe and love organic," Carl said, "but I am a numbers guy, and the numbers have to work. That's where we are right now, in the unknown zone. If you go by all the rules there is not a significant incentive to go organic at this point."

Right now, the farming industry is like a drug addict. Thanks to widespread and extensive use of synthetic chemicals, weeds and pests have, with time, built up a tolerance to the stuff. The short-term, profit-hungry practices of monocrop agriculture without cover crops or crop rotation and overtilling have only added to the problem, requiring more chemical use and making us—and our land—sick. If we continue on this path, the whole ecosystem will be in danger of an overdose. As USDA soil scientist researcher Rick Haney puts it, "We are essentially destroying the functionality of soil, so that you have to feed it more and more synthetic fertilizers just to keep growing this crop." Eliminating the drugs from the market is one way to solve the problem of our addiction—and that's what the organic certification has done. It's no doubt been a positive development, and yet a world without drugs that have the potential to alleviate pain and save lives would be an insult to the huge advances in medical technology humankind has made during our short stay on earth, regressing society back to the days when entire populations were wiped out by plagues. Carl's holistic model keeps the best of both worlds—no overdoses and no plagues, and a focus on soil health, which ensures no relapses.

Conventional and organic farmers alike are often placed in a frustrating conflict with their own land. If you choose the former, you put yourself, your workers, your land, and everyone who consumes your product at potential risk. Plus, the machine behind the marketplace separates you from knowing anything about your crop once it leaves your land. If you choose the latter, you take a financial risk, don't entirely solve the climate impact problem, and become wedded to a system that is open to corruption. There has been significant evidence that in India, the largest organic cotton exporter, some conventional cotton is being stamped as organic under lax inspections.

Even if properly regulated, the organic certification zeroes in on one end of the environmental concerns of modern-day agriculture: inputs. Yes, we need to pay attention to what we put on and in the soil, but what about soil health itself? Currently, there's no widely used certification or regulation looking just at soil health.

By embracing regenerative agricultural practices, Carl doesn't need to suffer from the demands of these two extremes. It also improves his own workplace happiness, if you will, because he gets to do what he loves most: farming. A holistic regenerative farmer has to actually have a relationship with the land, to pay attention to what his crops need, and to make decisions that come from experience and science, not just faith (although praying for rain never hurts). He'd still have to weed, but not as much, which would save him money overall. Taking things one step further, the marketing model of the cooperative could connect farmers to clothing brands with greater transparency; and if we make over the system the way this book aspires to, the brands will relay that information to their customers. In Carl's ideal world, "there would be a vertically integrated relationship between farm production methods to the consumer. Where there is transparency where everyone from top to bottom makes a good wage and the end user has a good product. And that transparency would deliver the economic stability that the farm needs, and it will give the consumer a unique connection and awareness of how their money spent is being used. There's a healthy balance of the total cost of what it takes to make a garment."

Though regenerative practices have significant benefits, they're no panacea. Some proponents of regenerative agriculture described it like a holy grail, claiming that these practices are the solution to climate change. Now, I hate to be the bearer of potential bad news, but the data on that is somewhat mixed. There is little debate that regenerative practices improve soil health and limit soil erosion, but the whole solving-climate-change thing is not perfectly established science. Proponents argue that we remove carbon dioxide out of the air by storing it as organic carbon in soils. But there is not a perfect scientific understanding of what keeps soil carbon sequestered; from that there is uncertainty whether regenerative practices actually sequester additional carbon in the long term. The truth is that more research is needed, as Rick Haney from the USDA underscores: "We need more independent research. We are only at the tip of the iceberg in terms of what we understand about how soil functions and its biology." It's unconscionable that the USDA research budget continues to get cut, and in its absence, voices with skin in the chemical game dominate the conversation.

Carl told me that, since my visit about a year and a half prior, he has seen a lot of interest and action from younger farmers in adopting regenerative practices. This is potentially very promising, because with an aging farm population, the next decade will bring a lot of new young farmers onto the land. Farmers National Company says 70 percent of farmland in the United States will transfer ownership over the next twenty years. This means we may well have more Carls on the land, open to tinkering and trying new things. Imagine the transformation we could see if we supported them with independent research to ensure that we understand and improve sustainable practices, and if we created financial incentives to nudge farmers toward climate-beneficial practices, breaking the cycle of more cotton, more chemicals, more clothes.

Two

Textiles Made in China: How the Drive for Cheap Is Killing the Planet

I t was like looking into a giant cotton candy machine. Except it wasn't candy, of course, but actual cotton fibers on their way to becoming yarn. At the other end of the factory, I squinted to see enormous rolls of dyed and finished fabric from which a garment— maybe a T-shirt, maybe a pair of jeans—would be made. The next stage in our jeans' journey is an intricate fabric of mechanical, political, economic, and environmental forces that allow the fashion industry to operate at its current massive, high-speed, global scale—for better or worse.

I was visiting the Zhejiang Qing Mao Weaving, Dyeing and Printing Co., located in Shaoxing, an industrial city of almost 5 million people located just south of Shanghai on the east coast of China. Inside this massive facility, raw cotton—the kind I had watched coming out of the gin in Texas 7,650 miles away—is further cleaned, untangled, homogenized, spun, woven, dyed, and finished into textiles ready to be cut and sewn.

How exactly did we—and our jeans—get from Texas to China? A lot of what was happening before me at Qing Mao used to happen far closer to home. The American farmers who came before Carl and Kent were more likely to send their crop following ginning to spinning and weaving mills in the States, as Carl told me, but those factories have all mostly shuttered thanks to laws and policies that have put company profits ahead of people and planet. As I stood amid the hum of enormous machines working at incredible speeds, I remembered the mill I visited in North Carolina, one of the few still in operation in the United States, a few years prior while sourcing for Zady. I pulled the image up on my phone to show the director of the factory floor, a slightly plump, middle-aged woman. Her eyes widened to almost fill her entire face and she broke out into a peal of staccato laughter. Her response, as relayed via my translator: "The machines are so old and short!"

Left: Spinning machines in a factory in North Carolina.
Right: Spinning machines in a factory in China.

While we may think of the United States as the home of everything cutting-edge, what little is left of America's textile production infrastructure is largely dated. In fact, the North Carolina spinning machine I'd photographed bore the name Saco-Lowell—formerly one of the largest textile machine manufacturers in the United States, but alas, no longer in business. I can only imagine how difficult maintenance is for machines when the company that made them is defunct.

If the US spinners were just old and hard to maintain, that would be one thing. But these older machines had not caught up with the times in many other ways that put them at a clear disadvantage when it comes to the global scale of production now required of textile mills. As my host pointed out, the US machines were short. Shorter machines have fewer spindles, which translates to smaller outputs of fabric. China, in stark contrast, has the latest, and biggest, machines, hence immense outputs. Qing Mao was just a small speck of the 73 million meters of fabric in the galaxy of clothing (and all things) manufacturing that happens all over China 24/7. Between July 2019 and July 2020, China produced 45.86 billion meters of fabric. That's enough fabric to wrap around the Earth more than 1,219 times. In 2015, China exported $284 billion worth of textiles and clothing and it commanded 43 percent of the global market. Growth has declined slightly in the sector since then due to a variety of circumstances, but in 2018, it still exported $119 billion worth of textiles that comprised 37.6 percent of the global market, far greater than the second-largest exporter, India, which has just 6 percent of the market. Hence its nickname, "the world's factory."

China's dominance in manufacturing today—and not just in the apparel industry—is the result of a confluence of decisions and policies that developed inside and outside the country. In the 1970s, China was reeling from the aftermath of the anticapitalist Great Leap Forward and Cultural Revolution, which together had completely decimated its economy and left tens of millions dead (some of us may not like where capitalism is right now, but China's attempts at a planned economy should at least be a warning that in life there are no panaceas). Determined to grow, the Chinese government began to move away from a planned economy and instituted some market-oriented reforms, including the 1978 open-door policy. Encouraging investments from foreign businesses to help boost Chinese exports, the Chinese market was proverbially "open" for expansion—although, it must be noted, the door did not open the other way to allow American companies looking to sell into the Chinese market.

Preceding China's open-door policy, major global powers post–World War II had themselves sought to begin opening their economies, based on the theory that countries that do business together are less likely to go to war with one another. Global trade existed well before this time—think the Silk Road and the cotton trade, huge international trade networks that were both based on clothing fibers—but it was a tiny fraction of most national economies. The new policy started to shift where things were being made. Wages in many countries overseas were far lower than those in the United States, partly due to the lack of worker protections (an issue we'll get into later in this chapter), and their output began to pose a threat to workers and manufacturers located in the States. By the 1960s, 5 percent of clothing Americans wore was being made overseas in the new emerging markets of Japan, Hong Kong (at that time under British rule), Pakistan, and India. By the 1970s, the number shot up to 25 percent. (And today, more than 98 percent of clothing Americans wear is made overseas.)

In response to the new and growing threat, a rare alliance was formed between American textile and apparel workers (through their unions) and their management to defend themselves from foreign competition. Together, they successfully lobbied for restrictions on imported textiles and apparel, which culminated in the 1974 Multifiber Arrangement (MFA), a single worldwide agreement based on a quota system. Under the MFA, the US and European governments developed a cap for each importing country on how many textiles/garments they could accept in an effort to protect their domestic industry.

While it was meant to protect against foreign competition, the MFA ultimately had the opposite effect. Instead of bringing manufacturing work that exceeded the quota for a country back to the United States or Europe, clothing company executives went in search of, and helped develop, low-wage manufacturers in other countries, like Bangladesh and Thailand, that had not yet had their quotas filled. The race to the bottom commenced.

An almost too-perfect case study of how all this rolled out is Levi's, one of the original clothing brands that defined American culture. Everyone from cowboys to hippies wore Levi's, and for most of its history, that meant that those cowboys and hippies were wearing American cotton made by American hands. From the time its riveted blue jeans were patented in 1873, Levi's sourced the cotton and labor for its products all in the US of A, which eventually helped crown El Paso, Texas, the denim capital of the world, churning out 2 million pairs of jeans a week in the 1990s, many of them Levi's. (We'll meet a key figure in Levi's empire, a wash-house owner in El Paso, in the next chapter.) At the peak of its presence in North America in 1996, Levi's employed 37,000 people (and was taking in $7 billion a year). Today, that manufacturing is gone. None of the company's jeans are manufactured in the United States anymore; almost none of the 450 million pairs sold in the United States can bear the signature gold tag "MADE IN USA."

But what's more is this: Levi's no longer really *makes* anything. While the brand name has stayed the same, behind the name, the entire business has completely changed. Levi's has transformed from a manufacturer into a merchant brand, the primary function of which is not to *produce*, as was the business model for clothing brands up until this point, but to design and execute on the best product assortment. This new business model—the merchant brand—was precisely what made the MFA go sour; since the model was less concerned with where and how things were made, CEOs closed their own manufacturing plants and were willing to drop longtime American manufacturers, like the kings and queens of denim in El Paso, for their less expensive counterparts overseas. We now see this business model reflected in the leadership of major fashion brands. Chip Bergh joined Levi's as CEO in 2011, after more than two decades in marketing and advertising at Procter & Gamble—not manufacturing.

What's wrong with companies focusing on marketing?, you may wonder. After all, if brands want to make the most money, they have to sell the most product with the greatest margin. But making the most money is *not* actually how or why the corporate form (the legal entity of most large companies) was created. It was first created to help pool capital to develop large projects for the *common* good—things like train tracks and hospitals. Academics call this a stakeholder model, since the public, the workers, and the owners (aka shareholders)—those who had stakes in the business—were all considered in how and what the corporation did. The whole model was founded upon our democratic ideals; the people elect the government that would provide charters to allow companies to be started for the people's benefit. The shift to being almost exclusively moneymaking machines began in earnest in the 1970s with the work of economist Milton Friedman. In breaking with the traditional stakeholder model, he stated, "There is one and only one social responsibility of business," and concluded: "increase profits." Shareholders, the people who technically own the company, thus took the seat of priority that all stakeholders once held.

This shareholder primacy model was itself part of a suite of ideas taking hold in business and government in the 1980s collectively referred to as "neoliberalism." Friedman was an intellectual godfather of the ideology, the heart of which is the idea that companies should be in the business of making as much money as possible and that unfettered markets (privatization, deregulation, hyperglobalization, free trade, reductions in government spending) lead to the best society. For a company like Levi's, and many others like it, it meant going from making jeans with some consideration for the workers in its factories and the community of El Paso, among others, to making jeans in a way that increased and maximized profit for the shareholders.

As the loophole of the MFA continued to incentivize overseas production and neoliberalist thinking dominated the economy, it became standard protocol for all companies looking to compete since the 1990s, in denim or anything else, to shift production away from the United States. Levi's had tried to stick to the stakeholder/manufacturing model, but found it impossible to compete in a marketplace full of shareholder/profit-maximizing merchant brands. Between 1993 and 2003, two hundred new brands carrying denim entered the market, and many were both trendier (because of the focus on merchandising) and lower priced (because of the manufacturing location). In El Paso, a pair of jeans might cost just under $7 to make, but just over the border in Mexico it's less than half that. In China, it's a mere $1.50 a pair. Levi's started feeling the pressure when revenue dropped from its peak of $7 billion in 1997 to $4.1 billion in just five years, never making more than $4.5 billion between 2001 and 2010. If Levi's wanted to improve on revenue, they must have felt they would have to follow their competitors' example and become a merchant brand to lower cost of production.

Cue the arrival of marketing savior Bergh, and many other executives in their respective roles, who began declaring things like: Levi's needs to "cut costs, drive cash flow, and become more data driven and financially disciplined to free up money to invest in technology and innovation." What does "cost cutting" and "financial discipline" translate to? Well, since the "what" of denim—namely, cotton—is already the lowest value-add component of denim production, that leaves the "who" (American workers) and "where" (America) up for grabs. It should be noted that executives, and their compensation, were never part of the "who" or "what" being cut; in 2002, then-CEO Philip Marineau's pay package was an enormous $25.1 million, a compensation that even exceeded Levi's entire net income.

===

Guess who was waiting with an open door (and millions of then quite desperate workers) to take up those cuts? Chinese businesses and the decidedly *not* neoliberal government that actively aided their efforts. And so the cotton for our "Made in USA" jeans started making the trek across the Pacific to be transformed into pants we could buy at increasingly lower prices, filling our brains with endorphins, our closets with stuff, and management and shareholders' pockets with cash.

It was no accident that China in particular was so ready, willing, and able to fill in the gaps of the MFA and merchant brands' desire for growth. The Chinese government had specifically selected the textile and clothing industry as a domestic industry to develop and promote under the open-door policy, since it already had the basic infrastructure and experience for apparel manufacturing from clothing its own people under the planned economy. Furthermore, as a very labor-intensive industry, it would also create a lot of jobs for the large and growing Chinese population. Plus, apparel did not require very advanced technologies, so it could get up and running rather quickly without needing a great deal of investment.

While the open-door policy liberalized trade, a feature of neoliberalism, China did not just let the market decide. The government was deeply involved in giving the garment industry favorable terms as one of its Six Priorities. This active government involvement in designing industrial policy created competitive advantages and allowed for the rapid expansion of the Chinese apparel industry.

Applying this model to other industries, sector by sector, China opened its manufacturing doors to the world and developed its economy at an incredible pace. The open-door policy culminated in China's joining the World Trade Organization (WTO) in 2001, which confirmed its status as a world economic power and spurred further growth of its garment industry. In 2002, garment output in China grew an additional 8 percent in just one year. The Western clothing industry was all but decimated at this point, and in 2005 the MFA ended, removing any quotas on countries' clothing exports to the developed world. The expiration of the MFA unleashed the Chinese apparel industry and allowed it to grow without limit.

And indeed it did. Country-wide, production in spinning mills increased at a jaw-dropping pace from 18 million spindles in the 1980s to

120 million spindles by 2015, or 48 percent of global production. El Paso lost its denim crown to Xintang, a town located in southeast China, north of Hong Kong, where, in 2010, 300 million pairs of jeans were produced. (In the summer of 2020, the crown was passed to yet another member of the fashion royal family—keep reading to find out who.) Xintang produces more than double what El Paso produced at its peak, 60 percent of all jeans made in China, and 40 percent of the jeans sold in the United States each year.

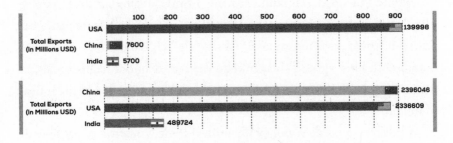

Compare the dramatic increase in China's exports versus
the United States in 1975 (above) and 2017 (below).

INSIDE THE WORLD'S FACTORY

Hence my arrival in China, to see for myself just how big the operations were in the textile factories that the world's factory comprises. Qing Mao, where I stood watching the giant cotton candy machine we opened this chapter with, is a premier Chinese factory. Its clients include many brands you have heard of, including Levi's and Abercrombie. The factory's history almost perfectly parallels the growth model of the Chinese textile industry. My guide, Charles Wang, the forty-something CEO of the company, had invited me to Qing Mao because he had nothing to hide. When he introduced himself, I was surprised not only that he came without an interpreter, but also that he spoke with a midwestern accent not unlike my own. He dressed in all black, and within the first five minutes of our conversation quoted both Nietzsche and Elon Musk.

He also had what appeared to be a world-class knowledge of tea and whiskey, a real Renaissance man. Charles was born in China and lived there when I visited with his family, but he was raised in Dearborn, Michigan (hence the accent), and spent most of his early career in the United States as a medical doctor and consultant at the prestigious firm McKinsey & Company.

Despite having reached what some consider the upper echelons of American success, Charles returned to China as an adult. You could say he followed his heart. His wife Yvonne's family owned Qing Mao, which they founded in Hong Kong in 1964 as a simple dye house. As the Chinese economy continued to open its doors, Qing Mao expanded into mainland China in 1996; following China's ascension to the WTO, they debuted the facility I stood inside in 2003 with expanded offerings of spinning, weaving, and dyeing. His father-in-law Jacob Wai had been aging all the while, so when the family started looking for a successor, they turned to Yvonne's new husband, whose American consulting background they thought could be useful. Dr. Wang, naturally, was up for the new challenge.

The first stop on the tour was a bit of déjà vu. Although it had been a month since I visited Texas, I felt the presence of Carl and Kent as I stood amid bales of tightly packed cotton, fresh off the boat and trains from gins in the United States and Australia. In the Qing Mao office, I smiled when I spotted a map of the cotton-producing US states, which included a big triangle for Lubbock. Qing Mao (and just about every other fabric mill) purchases bales of cotton on an exchange—that anonymous system that Carl had complained about—based on standard measurements for fiber length and strength, color, and trash percentage (more on that soon). I was told that the cotton being laid out on the day of my visit would be used to make fabric for a high-quality shirt for a popular mall brand you know. Workers unloaded the bales, then laid them adjacent to one another in large rows before they were slashed open by a serious-looking man wielding heavy-duty clippers.

Once all the bales were opened, a series of enormous machines, blowing a lot of air, opened and cleaned the cotton, extracting the bugs, dust, and remnants of cotton plants that the gins hadn't captured (that's the trash content referred to earlier). It was then tamed in a multistep process that reminded me of blow-drying curly hair. The machines reorient the fibers so that they face the same direction; these ropes of cotton fluff that I had seen earlier are called sliver (rhymes with "diver"). Slivers are wound up in great big cans to transport around the factory, where they might be combined with other materials like polyester, a fiber we'll talk more about later.

Then something really exciting happened. A friend joined us: a man on a special sliver-can moving vehicle, who chauffeured the cans to the combing station where short, low-quality fibers are removed. He was the rare human I saw on our journey through Qing Mao, whose person-to-machine ratio was surprisingly low. Clad in white aprons, face masks, and hair nets, the few humans we'd seen roved about like apparitions, overseeing the machines' work. Most of the employees, Charles explained to me, were middle-aged, due to the fact that factory work isn't exactly cool for young folks in China. This is causing a big problem in the long run for the country's growth-focused economy. Without a new generation to keep up the pace the industry might falter. "Everybody today is looking for the quick buck" in jobs like banking, he said, "but it's not sustainable."

Back to our cotton. The slivers still have a way to go before they can be spun into yarn—the steps of roving, spinning, and twisting all come next before we had something I recognized: yarn. But wait, there's more! Because we don't walk around wearing yarn from a belt at our hips and call them pants—that would be fringe, another look entirely—yarn needs to be woven into a roll of textile, which happens in three stages. Freshly spun yarn is "warped"—think back to first grade weaving—which sets up the vertical component of the weave, and "sized," in which starch is temporarily added to strengthen the yarn so it won't break during weaving, and then weaving commences. The machines that push and pull the yarn into a weave, like double-Dutch jump rope, operated at such great speed and corresponding noise that we had to enter the weaving room with ear plugs. All the while, inspectors were hovering. I saw glimpses of their blue uniforms peeking out from below the rows of machines.

From there we moved on to another huge room where the now woven fabric undergoes quality control (QC) to make sure there are no holes, pulls, random debris, or other flaws in the finished product. It is literally someone's job to stand all day and watch enormous sheets of backlit fabric roll through and note any irregularities. I recalled how I felt after

A QC worker at Qing Mao.

my last five-mile run—the hurts-so-good stiffness in my legs and knees and hips neutralized by the endorphin rush. This woman probably had lots of the stiffness, but none of the endorphins, after her shift was over.

After weaving, the textile is desized to remove the starch that was added before weaving. Scouring then removes excess oils and dirt; bleaching (not with a chlorine-based household bleach, but hydrogen peroxide) removes all remaining traces of the cotton's natural color; a sodium hydroxide bath improves the fabric's strength, flexibility, and feel; and in other rooms, the textile gets mercerized in another bath—a process that improves the internal structure of the fabric, its strength, flexibility, luster and feel, and absorption, and prepares it for dyeing.

I could spend an entire chapter just talking about dyes, but what you really need to know is that dyes, unlike many of the previous chemicals, are intended to stay on the cloth. In most of the previous steps, we washed (with water—the other "baths" all refer to chemical solutions that have some water in order to be liquid, but not exclusively) to remove excess chemicals from the textile—things like chlorine, which would yellow the fabric and give most of us a nasty burn if we wore chlorine-full clothes. All that washed-out stuff, called effluent, can have adverse impacts on the surrounding environment and community if it isn't properly treated. But dyes need to stick to the fibers, so they're resistant to fading from sweat (here's looking at you, athleisure), saliva, dry cleaning, light, and regular washing—a property called "color-fastness." Accordingly, the potential worrisome impact of dyes is in how they affect wearers. The dye is applied in a high concentration in water baths. In the dyeing room at Qing Mao, everything happened inside machines so the workers were fairly protected from the dyes they were handling, but as I'd see later on my trip, that's not always the case.

After dyeing comes finishing, where more chemicals are applied to make the textile wrinkle-free, waterproof, fire-retardant, or to give it other performance qualities. Shiny, soft, and available in a variety of colors, our fabric emerged from the entire process ready to be ogled by brands—and their customers.

HOT PANTS: CLOTHING, CLIMATE CHANGE, AND CHEMICAL POLLUTION

I did my share of ogling at Qing Mao, in addition to having done my fair share of walking, just following the fabric from start to finish. But more than impressed, I was exhausted—and frankly overwhelmed—by what a garment "Made in China" really means. The size, speed, and complexity of the operation was one thing, but what I couldn't get out of my head all during the tour was that each of these many steps requires an enormous amount of energy, and the combined load—just to make fabric—was monumental. Production in China is cheap in part because they use cheap energy. In China, 85.7 percent of the energy grid is fossil fuels (by comparison, it's around 63 percent in the United States), making the textiles that come out of Qing Mao (and the legions of other factories in China) not an insignificant reason why we have a global climate crisis. The country produces more carbon for the same amount of energy created than the US or the EU: 78 tons of CO_2 per unit of total primary energy supply, compared with the United States' 60 and the European Union's 54 tons. When you consider all of the clothing we buy, the number of energy-intensive steps to make the textiles of that clothing, and that China, which has such a carbon-intensive energy grid, is the leading producer of textiles, it's simple math to see how, according to one report, textile production creates more than 75 percent of the clothing industry's carbon footprint, and why the fashion industry more broadly is believed to contribute between 4 and 8.1 percent of the world's total carbon footprint. This, based even on the low estimate, means that fashion contributes the same level of greenhouse gases as France, Germany, and the United Kingdom combined.

If we do nothing about this and growth trends continue, by 2050, less than thirty years from now, one report estimates that clothes would use 26 percent of the world's entire carbon budget. By that same time, major global cities like Bangkok, Shanghai, and Mumbai will largely be under water during high tide. This data is terrifying, yet we're doing virtually nothing to stop it. Even the major "sustainable" standards and sustainably marketed brands in the clothing industry do not significantly address energy use in textile mills, which we now know is ground zero of energy emissions in the fashion industry. History shows us just how good we humans are at ignoring pending catastrophes—until, as recent history has shown, we're literally stopped in our tracks. US apparel imports from China decreased by 50.1 percent in July 2020 compared with the previous year due to the combination of trade tensions, the pandemic, and accusations of forced labor in cotton grown in China.

To be clear, China is not the sole party to blame for climate change. Hardly. The point here is to clarify why the fashion industry is such a massive player in carbon emissions—it's not that our jeans emit CO_2 into the air like the exhaust pipe on a Range Rover, but rather that lots of CO_2 is emitted to create the energy to make the jeans in the first place. People, including those in the environmental community, tend to think of our climate impact as things that take place within our own borders. But this is deceiving. As we see with our clothing, our carbon footprint extends far beyond our borders and must be taken into account when developing carbon reduction policies.

Our energy-sucking factories are not problematic for the environment just because they significantly contribute to climate change. Remember what they're used for: the application of chemical agents to make fibers bend (or straighten or stretch) to our human aesthetic and performance whims. You might not see them on your clothes, but they leave behind a trail that once you see—and feel, and smell—is pretty hard to forget.

A few days after my visit to Qing Mao, I landed in Guangdong, a coastal province in southeast China that is just opposite Hong Kong. Guangdong is one of the three largest garment and textile manufacturing bases in the country. It had sixty thousand factories in 2012 in the space of 150 miles. It is the same province that is home to Xintang, the denim capital.

Qing Mao is a five-star facility, but what I saw in Guangdong was perhaps more representative of the average Chinese factory in terms of how resources are used and cleaned up (or not) during the production process. I write "perhaps" because reliable data, especially when it comes to the environment, is really hard to come by in China. Access to physical locations is even harder. While I was actually invited to Qing Mao by Charles, I knew a less scrupulous facility would not exactly swing its doors open to foreigners, and certainly not foreigners writing books.

So I hired a fixer—he gave himself the title, I didn't. He is *the* man who builds relationships with businesses to allow a true "open door" policy with China—even when those doors need to be pried open. My fixer, whom I'll call Bruce, set up a meeting with a denim wholesaler, and I arrived posing as a denim buyer. Through a haze of cigarette smoke, the wholesaler informed me that his clients included Forever 21, which I believed based on the display of ripped acid wash jeans he showed off. Putting on my best denim-buyer game face, after asking about styling, turnaround times, and pricing, I asked how and where his jeans are made. Ordinarily, such a question would be cause for concern, but perhaps he took pity on me (I was seven months pregnant!). The guileless smile I pasted on probably also helped. The wholesaler led Bruce and me outside and down the street to the cutting room floor.

Standing side-by-side, we huddled under low-hanging overhead lights before several small mountains of jeans-in-the-making—single denim legs, or two legs halfway stitched together, waiting for a waistband and zipper. The placement of the piles certainly would have been in violation of any fire code, though something told me this place wasn't regularly inspected. That was my first tipoff that those cheap jeans were not being made under superb conditions. Betting my "innocence" could help me get more evidence, I asked, "That acid wash is so cool, how does it get that way?"

That happens at a different location at the wash house, he told me.

"That sounds awesome . . . could we maybe go see it?" I punctuated my question with my most winning smile.

Next thing I knew, I was in a black car being escorted by his company's driver to the wash house he used. From the backseat I could see that the driver was watching a movie and texting with three different phones all while dodging traffic. Fifteen minutes later, we arrived at an open two-story space. Even before I got out of the car, I noticed piles and piles of raw denim jeans strewn around on the ground; above them soft, finished jeans hung from hooks on the ceiling that spanned the entire building, like the biggest dry-cleaning conveyor I'd ever seen. Despite hurrying closely after my guide, I received many curious looks. They clearly didn't get many guests, certainly not foreign ones.

Luckily I had on rubber-soled shoes that day, because we crossed many inky puddles—the iridescent, bubbling contents of the industrial washing machines, which turn raw, stiff denim soft and wearable. Nothing was broken. Sloshing chemical runoff was the norm. The only clue to where it drained was a grate on the floor.

After acid washing (fun fact: jeans aren't acidified with true acid, it's bleach) comes finishing, and for this I was led to the side of the building to a small platform that jutted out onto the open air space of the wash house floor. I ascended the wooden staircase gingerly while the driver looked at me with concern as to whether it was a great idea for me, a pregnant woman, to keep going, noticing that there were gaping holes in the staircase and nothing below. At the top of the staircase, three people stood on the finishing platform (one wore a mask, none wore goggles), spraying a stream of pink liquid, probably the denim fading agent potassium permanganate. The European Chemical Agency classifies it as a "danger," as it has been found to affect the lungs if inhaled repeatedly, resulting in symptoms similar to bronchitis and pneumonia. Animal testing has also shown that repeated exposure to the substance causes possible toxicity to human reproduction or development. Needless to say, I did not stay there long.

I reunited with Bruce. I'd definitely seen even more than I expected to at this factory, but I knew there was more to the story—like where that blue muck and pink spray wound up downstream. We discreetly walked to the back of the factory where we saw a river, oil slicked and gray. Following the sound of a faint gurgling in the distance, we walked a little ways down the bank of the river. I knew that it wouldn't be good if anyone knew we were there, so I turned around every five seconds or so to make sure no one was following us. My travel mate, Alejandra, my colleague from New Standard Institute, took hold of my wrist and said, "Maxine, are you sure about this?" I shrugged off her comment, though a small knot inside my stomach tightened. Chemically ravaged rivers and sludge-filled factories are not exactly the images that the factories or the Chinese government is looking to project.

Making our way down a grassy, weed-strewn ravine, we came upon the source of the gurgling. A tucked-away pipe discreetly jutted into the river, out of which flowed a bluish-black substance that glistened like an oil spill. The air smelled foul; it burned my nose. We had found what appeared to be the other end of the pipe to the grate I had just seen in the wash house; the black pool spewing into the gray river was the effluents from everything that made those acid wash jeans look and feel the way they did. There was no sign that said "BEWARE: INDUSTRIAL WASTE!!" or flashing hazard lights to warn people of the contamination, not that it wasn't obvious. Nevertheless, a small elderly woman stood a few yards away, wearing a floral top, black cardigan, jeans, and ballet flats. She was bent down, farming. Farming along the banks of the black river.

At the end of my tour of Qing Mao, back in Shaoxing, we had stopped at their scrupulously run on-site water treatment plant. Charles spoke in detail about how the factory removes the chemicals used in production before the water is sent back to the river. I saw the process happen on a live monitor. But by the looks of the gray rivers in China, there are more factories operating like the one I visited in Guangdong than ones like Qing Mao. It's because of these unchecked effluents that, according to one China Water Risk estimate, the garment industry is the third-largest water polluter in the country, behind paper making and actual chemical production.

What exactly is in that murky water? I was about to find out.

"I AM SUPERMAN": WASTE REGULATION AND A HOPE FOR CHANGE

"And what's that beige part?" I was looking at a thick band that appeared along China's eastern shoreline on the country's equivalent of Google Earth. "We will go there tomorrow," said my companion, a very high-energy, jovial, twenty-five-year-young man named Xu Qingshi, but who goes by "Dashi," whom I'd been interviewing about his role in cleaning up China's waters. The band was polluted water, he explained, and the fact that it was so extensive it appeared with such clarity on the program was not good news at all.

At first glance, Dashi seemed to be plucked from a Silicon Valley lineup: hoodie-clad (screen-printed with Bart Simpson), running on Red Bull, and eager to pose for a selfie. He calls himself "Superman," which if he were a start-up kid would make me bristle in the way that Williamsburg hipsters do. (I live in Williamsburg, I'm allowed to say that). But in his case, though he said it with a laugh, I knew he was dead serious—and he is just the kind of superhero our rivers (and oceans and air) need.

Dashi is the founder of Zhaolu Environmental Protection Center, a small but mighty environmental organization operating out of a scrappy ground-floor space in Shaoxing, the same city where Qing Mao is located. After observing many people in his community getting sick with cancer and asthma, he connected the dots to the black and acrid waterways that surrounded him. He decided he needed to do something to clean up his home. He had also seen the movie *Erin Brockovich*—starring

Julia Roberts as a single mom turned environmental activist—and took away from that film that regular people need to stand up in the face of industrial malfeasance and government inaction. (Julia Roberts herself may have had something to do with it. While we had spoken strictly through an interpreter for the two days we were together, in telling about his inspiration, Dashi pointed to a picture and said, in clear-as-day English, with a massive, childlike grin on his face, "Julia Roberts, very beautiful.") I asked him why he focuses on chemical pollution and not air pollution and climate change emissions. He threw his hands up in the air and said, "Because it's much harder to test the air."

At Zhaolu, he and his team of eight engineers hunt for chemical villains by testing water samples near factories. They then hand that data over to the government, which then, hopefully, takes corrective action against offenders. I write "hopefully" because China has, since the open-door policy, adopted a growth-at-all-costs model. While there is evidence that the government has imposed some corrective actions against rule-breaking companies, even forcing some company closures, it has had only a limited effect. Corrective actions can increase costs for companies, which can make them less competitive in a global market that does not put a cost on pollution. (Before we judge China too much, we should remember that the United States and Europe also had this model during industrialization, and the US, through the Trump administration, was clearly sliding backward in this regard.) While there has been enormous economic growth, pollution has taken a devastating toll in China, including an estimated 1.2 million premature deaths from air pollution. The health impacts are so high that some experts believe that the pollution threatens the Chinese government; incensed citizens might take drastic actions against the powers that be if they continue to die for the sake of growth. This leaves the Chinese government in a complicated situation, trying to simultaneously hold on to power by maintaining the stratospheric pace of economic growth and promising to curb pollution, two paths that don't easily lead to the same destination.

So while I did hear anecdotally during my conversations in China of factories getting shut down for violating emissions limits, which contributed to reducing particle air pollution by up to one third across industrial cities in China, I also heard that there was a strong correlation between government punishment of offenders and the country's economy. Anecdotal evidence is about as good as it gets; there is little reliable data available to the public. Regulations were more strictly imposed in good times, but as soon as the pressure on the national economy increases (as it is today), enforcement loses priority. All of this leaves Dashi and Zhaolu in a precarious situation. While he introduced his work by showing a brief video about the center, Dashi stated rather casually, "Oh, my friend," referring to a man onscreen, "he is in jail now." Dashi ostensibly helps the government with policing, but in the end he's just another whistle-blower, and therefore always at risk of being detained

Chemical dumping in Guangdong, one of the three largest garment and textile manufacturing bases in China.

and imprisoned himself should the local government decide the environmental work is too disruptive to potential growth.

Hearing the desperation of Dashi's plight is frustrating in and of itself, but even more so because China's current pollution problems were entirely preventable. America's cities were thick with smog, its rivers flooded with muck and mire just a few decades ago, and thanks to citizens like Dashi, our local and federal governments left a rather clear blueprint for how *not* to let that happen again. Compare the haze of Shaoxing today with the haze of New York City in the sixties.

The haze over New York in the sixties (left)
versus the haze of Shaoxing today (right)

It was a grassroots movement in the United States protesting this dangerous pollution that led to the Clean Water Act (1972) and the Environmental Protection Agency (EPA), which created laws and an enforcement mechanism to limit pollution from the textile and energy industries. And yet, when American companies moved their production to China (and elsewhere), these protections were disregarded. Governments seeking to attract Western businesses were disincentivized to create the same costly environmental protections. So while we no longer see this type of pollution in our backyards, and hear fewer Erin Brockovich–esque reports of cancer pods in suburban America, we shouldn't boast about lowering US pollutants. We just shipped them halfway across the world (and for the pollution that does still happen here, we tuck that away in poorer and minority communities). This wasn't an inevitability of globalization. We could put language in our trade deals that insists on *at least* the same environmental and social standards that we have in the United States; we just haven't—yet.

The rivers I visited in the textile cities of China were "black and smelly" (a qualifier that an environmental group officially uses when reporting different kinds of water pollution) and the air around those factories was difficult to breathe. I could guess what kinds of sickness Dashi observed in his community from the minute I stepped off the train in Shaoxing, when my eyes almost immediately began to sting. The next day, waking up in a luxurious bed on the thirty-sixth floor of a slick new hotel, I was greeted by a smog-filtered sunrise and soreness in my eyes, nose, and throat. Even before I saw the workings of Qing Mao, I drove through streets of Shaoxing overcrowded with cars, motorbikes, and trucks of all sizes. Interspersed with the traffic was the occasional old woman selling vegetables, perhaps like the lady I saw by the river in her neighborhood in Guangdong; and just beyond, through that same thick layer of smog, the banks of rivers the color of oil spill.

The effluents I could see from that precarious riverbank (because there were certainly pollutants spewing from that pipe I could *not* see) were almost certainly from dyes—which by nature of their color makes them hardest to cover up, even by the slyest of factories. While dye-filled rivers were behind the rise of the EPA in the United States, as well as previous initiatives by Greenpeace to tackle water pollution in China, environmental toxicologist and former Natural Resources Defense Council (NRDC) researcher Linda Greer believes that dyes today are actually less of a problem than they used to be. Most of the worst offenders have been banned for decades, since the first wave of workplace disease in the 1950s. It's still not great that they end up in rivers, but it's at least why dye-related illnesses are rarely reported by consumers. And though dyes are not as bad for our skin as they once were, they do still contribute to energy demands and thus carbon emission. Cotton, for example, is a fiber with a low color-fastness, and therefore requires lots of dye plus upwards of ten hot-water baths to ensure that it won't run or fade when you wear or wash the garment.

And then there's all those invisible chemicals that are used throughout the textile creation, dyeing, and finishing processes. The German Catalogue of Textile Auxiliaries 2008–9 report found 5,800 agents in textiles containing four hundred to six hundred ingredients—and that's not including dyes. Final fabrics can be comprised of 28 percent chemicals by weight—and that's only if they're made of 100 percent *natural* fibers; synthetic fibers have even greater concentrations. As I saw in Guangdong, and as Dashi showed me in Shaoxing, depending on the existence and quality of a factory's wastewater processing system, a little or a lot of those chemicals will wind up in local rivers, streams, or groundwater, and in aquatic ecosystems. Because they're transparent, unlike dyes, they're harder to detect in wastewater; you might look at a river full of polyvinyl acetate (PVA) or formaldehyde and think it's safe to drink from or use to water your crops.

Freshwater (i.e., not from the sea) is a highly limited commodity and its supply is diminishing. In fact, a full two thirds of the world's population are projected to face water scarcity by 2025, according to the UN. This is not because water molecules are disappearing; it's because existing freshwater is getting polluted, rendering it unsafe for human consumption, and the chemicals from our clothing contribute to this water crisis. It's happening in Shaoxing, where average daily discharge of printing and dyeing wastewater exceeds 600,000 tons. And it's spreading elsewhere along the garment production map, including to Bangladesh and Cambodia, where water for humans, livestock, and agriculture is increasingly toxic.

How extensive is the problem? Well, we looked at factories just dumping their effluents into the rivers, and the rivers I saw in China were mostly all very dark gray. The ones in Bangladesh, where we'll head in the next chapter, were even worse. If you Google something like "textile pollution + water," you'll likely come across the statistic that the textile industry is the second-largest polluter of freshwater in the world, or that 20 percent of industrial water pollution comes from the apparel industry. But do some further digging, as I have for you, and you quickly come to a dead end. There is actually no primary resource for those claims. It's the fashion industry's version of fake news. Although it could be true, we just don't really know. It's frustrating, and far less tweetable, but there it is. We need research commissioned on these things so we can actually grasp the enormity of the problem and track its sources in order to even begin to address it.

What we can do is pinpoint some of the harm that effluents cause. Sulfur, naphthol, vat dyes, nitrates, acetic acid, soaps, chromium compounds; heavy metals like copper, arsenic, lead, cadmium, mercury, nickel, and cobalt; and auxiliary chemicals render effluent dangerously toxic. According to KEMI, the Swedish chemicals agency, approximately 10 percent of 2,400 identified textile-related substances are considered to be of potential risk to human health. Polluted water has been found to cause things like painful skin disease, diarrhea, food poisoning, and gastrointestinal problems in the short term; and extremely serious health implications such as respiratory problems, cancer, mutagenicity, and long-term reproductive toxicity in the long term.

The first way chemicals affect health is through the food chain. The woman I spotted at the river in Guangdong is not the only person on this planet farming with effluent-tainted water. Studies have demonstrated, not surprisingly, that vegetables collected from these farms show the presence of textile dyes, which people then ingest when they eat those vegetables. The fish who swim in those waters, which are also eaten, can also be contaminated.

Even the entirely nontoxic material that makes its way into the effluents can have devastating consequences, the most common among them being starch. When too much starch enters the water from insufficient treatment, the starch depletes the oxygen in the river and kills fish.

*Agricultural plots relying on an effluent-filled river in Shaoxing, China.
Note the factory across the river.*

Even leaving aside chemicals in the water, there's also the chemicals that don't get washed off our clothes. Since skin is *not* an impenetrable barrier, chemicals that rub up against our bodies all day can and do enter our system through the skin. Consider this one study from 1978 that was looking at children's pajamas that were known to contain a now banned chemical flame retardant that causes cancer. While the chemical is no longer used, the same biological process of absorption through skin can occur with other chemicals. The children wearing those pajamas were found after one night of sleep to have a 5,000 percent increase in concentration of a by-product of that substance in their urine. After changing to pajamas free of the substance, the concentration of the substance decreased, but after five days, it was still twenty times higher compared with the initial concentration. There is an even greater risk of chemical uptake when we sweat.

Thus far, researchers have focused most of their attention on the effects of chemical exposure on textile workers. From that, we know

conclusively that some chemicals used in the garment industry may expose workers to significant health risks, including illnesses like bladder
cancer. For textile wearers—that is all of us—the chemicals we know to
be dangerous per laboratory studies exist in only trace amounts once
they get to us; chemicals that get washed off at the mills are less con-
cerning for the wearer than chemicals in performance finishes that are
meant to stay on the fabric. Unfortunately, it's no surprise that we don't
know the whole story about a lot of things, including these chemicals'
true impact: What are all the varieties and quantities of chemicals on
our clothes? And what are their health effects in the concentrations we
may be exposed to when we wear them?

If you check out the label on your clothes, it won't tell you what
chemicals were used to make them. As we explored in chapter 1, not even
the organic standard helps us much here when it comes to these chemicals. If a garment says it's made from 100 percent organic cotton, it only
means the cotton was *grown* without the prohibited pesticides, synthetic
fertilizers, etc. "Organic" has nothing to do with the chemical processes
that we just went through in great detail in the textile creation phase.
Beware of brands that overpromise and equate "organic" with nontoxic,
clean, or safe.

Given the available data, and given the *potential* risks and the sheer
volume of chemicals used in making garments, we need additional research on the impact to wearers. Without further research, and enforced
regulations to ensure proper chemical use and management at our textile mills, the workers, the communities that rely on the water that comes
into contact with effluents, and to some extent we the wearers, are exposed. And while we have a few amazing superheroes on the ground like
Dashi and his selfless team, we need more people, whose work does not
put them at risk of persecution, to help us remake the world so we all
wake up to clear skies and clean water.

"THE FUTURE IS PLASTICS":
SYNTHETIC FIBERS, CLIMATE CHANGE,
MICROPLASTICS, AND YOU

We've spent a lot of time going over what's not on our jeans' tags: chemical fertilizers and pesticides, the carbon emissions and more chemicals used to make and finish the fabric itself. But there's another hidden ingredient in our favorite pants: fossil fuels.

While you may not know that fossil fuels are a key ingredient in your clothes, you may be familiar with some of their pseudonyms: polyester, nylon, elastane (better known by its brand names Spandex and Lycra), and acrylic. These are all a kind of synthetic fabric made of plastic, which itself is mostly derived from fossil fuels. Without synthetics we would not have skinny jeans, half-zip fleece, yoga pants, vegan "wool" scarves or "leather" bags, or much of what we know as fast fashion. I'm going to focus on polyester, because it is by far the most common synthetic fiber; in fact, as we'll see, it is today the most common fiber, period.

The origins of polyester read like the story of penicillin or the lightbulb—a miracle of human innovation and mastery over nature. Born of the same wartime and postwar research behind chemical pesticides and fertilizers that launched an agricultural revolution, polyester became the poster child for modern fashion and modernity writ large in the 1950s. It was in fact a scientist at the chemical giant DuPont, Wallace H. Carothers, who discovered that long chains of molecules of repeating molecular units of atoms, or polymers, could be combined in a chemical lab to form a synthetic fiber. Celebrated for their man-made uniformity and adaptability, synthetics could be shaped into many different textures and types of textiles that cotton or silk, with their fixed properties and susceptibility to the whims of nature, could not. Fossil fuels entered the scene in 1941, when Carothers's early work in polyester was advanced by two British scientists, John Whinfield and James Dickson, who patented polyethylene terephthalate (PET or PETE), which is the chemical name for polyester. DuPont bought the rights to this technology in 1946, and started developing its own formulas in 1950.

The fossil fuel–based PET can be used to make many things, from soda bottles to the clothing on our backs. When melted, PET has the consistency of lukewarm honey. If you squeeze it through a device called a spinneret, which looks kind of like the shower head in your bathroom or a pasta maker, you get long, continuous filaments that are drawn together to become yarn. Weave that yarn together, and you have a fabric.

Polyester and its fossil fueled–based brethren have completely, and silently, transformed how we dress ourselves. We went from closets full of things that came from farms and ranches (cotton, linen, wool) to closets full of fossil fuels—plastic—in a single generation. In fact, the most accurate depiction of clothing's origins today would not be the sun rising on a cotton farm, but an oil rig drilling into the ocean floor.

While the United States still produces cotton fibers thanks in part to US farm subsidies, it is not a global competitor in the polyester fiber market. As with textiles, China has also come to dominate polyester fiber production; it produces well over half of the global supply. In fact,

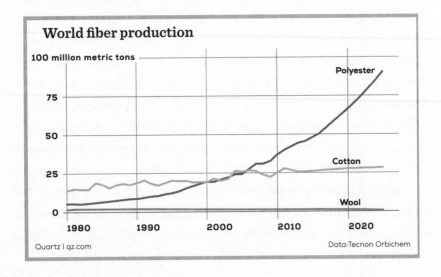

World fiber production
100 million metric tons

Quartz I qz.com   Data:Tecnon Orbichem

Zhejiang province, where Qing Mao and Dashi are based, is also home to several of the world's most significant polyester fiber companies.

Why does the explosive growth of these fossil fuel–based plastic fabrics matter? Two reasons: First, it contributes to climate change. As we now know, most polyester starts its life as crude oil, a fossil fuel. The fiber production stage—growing cotton or drilling oil, and converting that oil to polyester fiber, but not spinning and weaving—represents 15 percent of the total carbon footprint of our clothes, but polyester has a larger footprint than natural fibers.

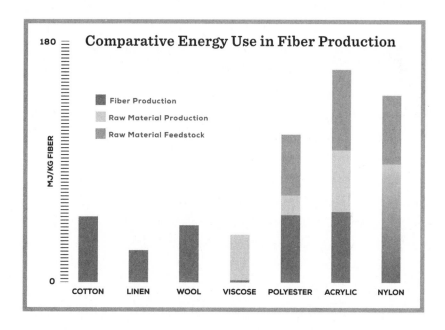

Second, polyester contributes significantly to plastic pollution. When we wash our plastic-based clothes, whether it's during the production process at garment factories or in the laundry at home, some of the polyester sheds off into the water in the form of microplastics that can get past even the best water treatment methods. These tiny shreds of processed crude oil have made their way to the tops of mountains, to the bottom of the ocean floor, and everywhere in between. By 2050 there will be more plastic than fish in the ocean, by weight. You might assume that that plastic is mostly straws, shopping bags, or water bottles. But a great part of the plastic crisis is largely invisible—only around 6 percent of the total mass of plastic entering the oceans can be seen by the eye.

The issue of plastic in the ocean is only relatively recently recognized— and researchers are still wrapping their collective heads around it. There is much to be learned about which of our clothes produces the most of these fibers, and what exactly they do to the ecosystem and to us humans. There is no scientific consensus on the percentage of global plastic production that goes into textiles, nor do we know the proportion of plastic pollution that comes from our clothing. However, the totality of what *is* known and estimated is still cause for serious concern.

We know that synthetic microfibers have been found on every corner of our world, from the beaches of Florida to Arctic sea ice, in the air we breathe, the salt we eat, the bottled water we drink, and on the fields that grow our food. The first study ever to research microfibers released from clothing during a laundry cycle found a single garment could release more than 1,900 microfibers. A subsequent study found more than 700,000 released in a single load of laundry. Consistently, synthetics released more microfibers than natural fabrics. As much as 209,000 tons of synthetic microfibers enter the marine environment in a single year. As these fibers enter the oceans, animals are confusing these fibers for food, so on it goes up the food chain until we end up consuming it, too. In one study, a quarter of fish in a California fish market were found to have synthetic microfibers in their systems. Microfiber-eating fish have been found to reproduce less, and their offspring, even if they were not exposed to plastic particles, also have been found to have fewer offspring—a big ripple effect in the marine ecosystems. Remember, plastic was beloved because it sticks around forever. I suppose DuPont didn't think that one all the way through.

While plastic, "miracle" fossil fuel product that it is, still plays an important role in hospitals and for things like food safety and sanitation (and I'll even concede for certain clothing uses, like stretchy gym clothes), it doesn't belong everywhere. But it *is* everywhere, literally, which is a rallying cry to the fashion industry and us, the drivers of this consumer industry, to start seriously questioning whether these unknown risks are worth it for something like a shirt you might not even like when you buy it, let alone tomorrow.

———

As my airplane took off from the Hong Kong airport, the gauzy clouds were already occluding China's poisoned shorelines. I was no longer inhaling the stomach-churning aroma of black and smelly water, but I liberally squirted drops into my eyes and sucked on throat lozenges to soothe them. Below me was a global superpower that earned its cape in significant part thanks to the clothing industry. China's modernization via the textile and apparel industry has brought millions of people out of poverty, and has continued to connect us in some ways financially and socially. Fifty years ago, it'd be inconceivable to think that American trends would show up on the banks of a secluded river in China, as my friend the farmer revealed to me in her ballet flats and cardigan.

And yet clothes are also behind the dissemination of technologies that put us on the brink of a global crisis. Thanks to coal-powered manufacturing, unregulated chemical management, cheap synthetic fibers, and an insatiable appetite for more, our planet and our bodies (both textile workers' and consumers', but principally the former) are being exposed to things they shouldn't be, and depleted of what they need to survive.

The places where climate change and pollution originate are becoming harder and harder to see—literally, for me, as clouds obscured the country below me, but also metaphorically as Westerners punt responsibility for manufacturing across the world. And as the Chinese government looks to move away from this polluting industry and as the industry continues to look for cheaper sources of production, other countries like Bangladesh, Vietnam, Cambodia, and Ethiopia are quickly entering the scene. These developing nations are on a fast track to repeating China's fate if we don't change course, and by the time they reach China's levels of growth it will be too late. How can the rest of the world, which is not regularly exposed to those toxins and doesn't make the connection between their clothing and climate change, begin to care?

But then I remembered Dashi's enthusiastic promise: "I'm Superman," he said over and over, and so is every person reading this book. Whether we admit it to ourselves or not, the clothing we choose to purchase has an impact, so it's up to us to demand that the industry behind what we wear and our government, which sets the rules of trade, get out of the way in allowing the planet and its people to thrive.

Three

My Factory Is a Cage:
Cut and Sew and the Crisis of Labor

I arrived in Dhaka, Bangladesh, on a steamy May morning. The city assaulted my senses: the aural cacophony of human voices, car horns, train noise, and street cart wheels navigating the obstacle course of potholes, rocks, and debris along the roads. Busy marketplaces where bodega-like stands sell everything from oranges to motor oil, and the colorful sea of saris blend high and low fashion with unstudied ease, contrasted sharply with peaceful, barricaded enclaves protecting beautiful, modern concrete-and-wood architecture set within man-made parks, seemingly torn from the pages of a design magazine. Buses, cars, vans, carts, bikes, rickshaws—so many rickshaws and trucks executed dizzying choreography, resulting in impossible traffic that makes LA seem like a 1950s suburb.

The heat and humidity intensified everything, the way the sweat dripping off a glass of water onto a magazine makes the colors bleed and deepen. Smog and pollution hung in the air day and night like honey. Just a few hours after arriving in Dhaka, the smog had completely clogged my nose and throat, so that my hacking cough became part of my personal soundtrack of the city.

Although several months (and one child) passed between my trips to

China and Bangladesh, this smog erased time, connecting the dots in my senses between these two stops on our jeans' journey. You'll recall from our time in China that the role of the fashion industry in that economy is at a turning point. In the last decade, a conflict has arisen in China's factories between growing demand for product at a faster pace and even cheaper costs and the people doing all the cutting and sewing to make it happen. As the Chinese economy has expanded to more advanced, higher-wage industries, Chinese workers seem to have more of a choice. And they're not choosing garment work. With price as the number-one driver behind any decision, the fashion industry likewise has started sending cut-and-sew business—the part of the process that requires the most labor—to other countries with even lower wages.

In pursuit of winning this race to the bottom, the fashion industry—particularly the cheaper brands, though not exclusively—has found Bangladesh (and other countries such as Sri Lanka, Vietnam, Cambodia, and Ethiopia). In a 2017 McKinsey survey, 62 percent of chief purchasing officers at US and EU apparel companies said they were planning to diversify their sourcing away from China in the near- to medium-term. Bangladesh, Ethiopia, Myanmar, and Vietnam emerged as the top countries where respondents expect to increase sourcing. In 2018, China was still the leading garment exporter at $152 billion, and Bangladesh a distant second at $34 billion. But we've closed that gap in a remarkably short time. As I teased you in the previous chapter, the title of top denim exporter to the United States was recently taken away from China by—guess who? Bangladesh.

Wedged in the crook of the Bay of Bengal between India and Myanmar, Bangladesh is a country the size of Iowa but home to 163 million people, about half as many people as in all fifty United States combined—making it the eighth most populous, and one of the most densely populated, countries in the world. Its capital, Dhaka, was once known as the "Venice of the East" because of all the rivers that run through it—an image that sticks even today as gondola-like boats and massive cargo ships share the waters surrounding and cutting through it (if you just block out the litter that engulfs the rivers). Back then, it was a capital for cotton and silk; today, Bangladesh remains increasingly integral to global clothing production but, as we'll see, more for its abundant human resources than its natural ones.

The chaos of the cramped capital city is a reflection of Bangladesh's rapid growth over the last decade, which makes it one of the fastest-growing economies in the world. Because 80 percent of the country is located on a floodplain, there has been a massive influx of people to the cities trying to escape the soggy countryside, where the viability of the traditional agricultural livelihood is decreasing. Because of its geography, Bangladesh is also one of the world's most vulnerable countries to climate change. One should not lose sight of the irony that this country, which is so invested in such a significant contributor to climate change, is also suffering the most from its extreme weather.

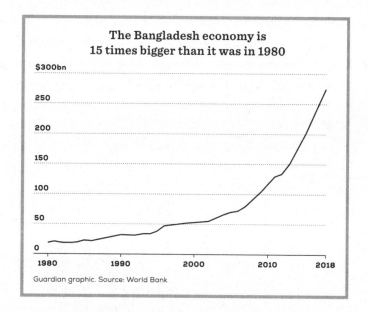

The Bangladesh economy is 15 times bigger than it was in 1980

Guardian graphic. Source: World Bank

This new population of eager workers has contributed to 188 percent expansion in GDP since 2009. Thanks to this boon, the population below the poverty line has fallen precipitously, landing at 14.5 percent in

2019, from more than 44 percent in 1991. The incidence of neonatal and maternal mortality has shrunk as well, and the average lifespan has increased by an unrivaled one third compared to the 1980s.

If moving garment production to Bangladesh was the tide that lifts all boats, then the shift would be excellent news all around. But, of course, that's not the whole story. Not according to what I saw, and not according to what the more nuanced data reveals. First, all that economic growth is not equally shared. As I walked through Dhaka and its environs, the rift between rich and poor couldn't be more apparent. Teeming slums and urban marketplaces are mere steps away (though hidden) from those cordoned-off enclaves of luxury, the juxtaposition even more jarring to a foreigner like me for whom both worlds are so different from my own.

And while incomes and lifespans have increased, there is an upper limit to that progress if the industry continues to stay focused exclusively on price. We can look at wages to see this at play. In 2013, the average nominal wage for garment and textile workers in China was $491 per month, while their Bangladeshi equivalents earned $163 per month. Several countries now on companies' hot lists for new areas of expansion ranked even below this figure, such as Ethiopia, which has a

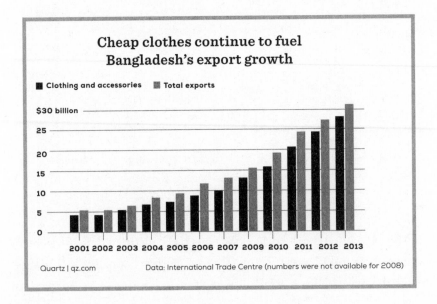

Cheap clothes continue to fuel Bangladesh's export growth

■ Clothing and accessories   ■ Total exports

Quartz | qz.com   Data: International Trade Centre (numbers were not available for 2008)

$26 per month salary. Bangladesh and other developing countries have been invited to the apparel party so brands can continue their race to the bottom. When Bangladeshi workers are deemed too expensive, manufacturing will go elsewhere even as companies turn a very blind eye to the repercussions of that race.

Strewn like technicolor leaves in the waterways separating a major slum, Korail, from its high-rise neighbor, were scraps of clothes—alongside plastic bags, rotting wood, rubber hoses, and every other variety of trash. Here was the residue of the fashion industry that has helped modernize Bangladesh; the industry now comprises 12.26 percent of the GDP and 83.9 percent of total exports. But at the same time, as we will see, the industry threatens to rip it apart at its already straining seams. The question now remains: Will increasing investment in apparel ultimately hinder Bangladesh's development, or can our clothes be the path to upward mobility and long-term prosperity for this country, the whole industry, and the broader world?

The back of the Korail slum, with a river of fabric scraps and other trash separating it from the wealthy neighborhood in the distance.

CUT AND SEW: TAKING HOME EC TO SCALE

A giant white SUV arrived at my hotel for my tour of one of the outposts of a respected, large-volume factory located in Gazipur District, outside of Dhaka, an area that has become popular for garment production since the sector has formalized and largely moved out of the city center. Climbing into the car, I couldn't help but sigh. Here I was, wanting to get a real "inside scoop" of factory life, and my commute—as generous as it was, organized by the factory—couldn't be further from what a worker experiences.

But within an hour (of the one and a half hours it would take to travel the sixteen miles), it became clear that even inside the steel cage of the SUV, I could still get a glimpse of life outside the factory walls. My driver and the CEO of the factory's assistant were chatting away in Bengali in the front seat, while my face was plastered to the window. We had taken a sharp turn out of the city into another world. The eclectic mix of greenery, abstract cement sculpture, and traffic suggested that Dhaka's rising metropolis had been replaced by another kind of chaos. Stretching out before me was a part-rural and part-industrial wasteland.

We bumped along the dirt "road," or the place where the cars were, avoiding pedestrians, rail-thin rickshaw drivers, and potholes in equal numbers. Beside the road was an empty railroad track littered with garbage and broken pipes. Like rainbow sprinkles on a chocolate birthday cake, the mounds of mud were dotted with colorful bits of trash: cartons, plastic bags, mismatched flip-flops, and, of course, fabric scrap. Fabric scrap everywhere. A few audacious papaya and palm trees stood their ground amid the refuse of man's material desires, but not all had won the battle—enormous tree trunks and upended root systems lay like fallen soldiers on the front lines of man's war versus nature.

Farther in the distance, inky black rivers cut through haphazard agricultural fields—another casualty of the war. The landscape transported me yet again back to China, where Dashi had unsuccessfully tried to convince me that a stinky black river was, in fact, "clean." I'd never seen water so polluted at the time, but now I suddenly understood what he meant. *This* is what a black river looks like. Three of Bangladesh's rivers are now "biologically dead," supporting no forms of life, due to pollution from the apparel and leather industries.

Gradually, the amount of fabric scrap increased. We were getting closer to the factory. Driving through this more industrial area, I looked inside some open buildings—one was occupied solely by a mountain of fabric scrap, another by endless rolls of fabric waiting to become garments. The inputs and outputs of what we wear were piled high, spilling onto the street and mixing with mud. I was inside the proverbial sausage factory.

At the Qing Mao facilities in China, raw fibers like cotton or polyester were transformed into a roll of textile. Here in Gazipur, I entered the extraordinary world of cut and sew, where those rolls of fabric become finished garments. I had been awed by the scale of operations at Qing Mao and the comparable filth of the rivers that surround the factories, but here I saw another kind of detail, skill, and efficiency that both awed and disgusted me. China may be the world's factory, but the workers on the cutting and sewing floors are machines themselves, producing garments at an unthinkable pace. The question remains: Are humans meant to be machines?

The factory that was my destination had more than ten thousand employees stationed on seventy production lines in two locations. Fabric from China, India, Turkey, Vietnam, and Pakistan enters; the factory spits it out in the form of over one million garments a month. Those clothes make their way into stores and onto websites for Gap, Ann Taylor, and American Eagle, among others. Looking out over the sea of people working these lines, I recalled some of the impulse buys I'd made back in my law school days. Did I ever even stop to think that a person—maybe one in this room—had put together the item I bought and tossed as casually as my plastic-coated paper cup of coffee?

Unfortunately, I couldn't get access to a denim factory in Bangladesh. Although this factory didn't make jeans, I did watch two kinds of garments being made from start to finish: a women's cargo jacket and a pair of men's trousers in a process not dissimilar to what would have happened to make jeans. Step one was cutting, in which several pieces of fabric were stacked on top of one another, with a paper pattern on top. On the back wall hung a rack of patterns in every shape and size imaginable— half of a sleeve or pant leg, a collar, a cuff, a back pocket—which, when eventually sewn together, would make a complete garment. Each individual segment gets its own pattern, digitally delivered by the design team of the brand; the segments themselves are cut out of long rolls of paper by a machine in a separate area. A big saw—like the jigsaw from shop class—traces the pattern to cut through the layers of fabric. Also like shop class, it takes quite a degree of skill to follow the pattern precisely through all the layers without creating jagged edges—or, more important, without chopping off one's finger. Hence why the man operating the machine was wearing a metal glove, like a knight's chainmail, to protect against such accidents. Because cutting involves larger machinery, it's been deemed work more appropriate for men; so unlike everywhere else in the cut-and-sew process, this floor was half Y-chromosome.

From there, all those individual cut pieces are brought upstairs for the production lines to begin their work of assembling the garments. As I accompanied the legs and sleeves and collars and lapels up the dimly lit stairwell, I caught sight of a set of three small posters hanging high up on the wall, well above my line of vision. I could make out a few of the small, densely printed words in English: the names of brands and the words "code of conduct." I had seen big, colorful posters about workers' rights and minimum wage laws in offices where I worked at home— they were always placed in prominent, if not aesthetically pleasing, locations, like the kitchen area or copy room. Could it be that this was the only notice of workers' rights in the whole building? I didn't have time to linger on the posters or ask about them—there were clothes to be made—but the image stuck with me.

The production—or sewing—floor was buzzing, literally and metaphorically. It echoed endlessly into the distance over and around row after row of sewing machines and workstations manned by, well, not men. Sewing is women's work, as it has been since preindustrial and early industrial times. While it is very difficult—maddeningly so, in fact—to find reliable, precise figures on the percentage of women in the industry worldwide, by all accounts and in my own observations, it is largely women who power the machines that make our clothes. A figure that 80 percent of cut-and-sew workers are women circulates, but has not been substantiated.

Upon my arrival on the floor, I felt my body tightening, responding to the tension in the atmosphere. Looking around, I noticed gauges of speed and productivity displayed everywhere. The factory floor was organized in twenty-two long rows. At the front of each row stood a large whiteboard with a list of the production targets and actualities, broken down by month, day, hour, and half-hour. In one row of sixty-six people (fifty sewing machine operators, ten helpers, and six ironers) on that day, 1,600 pairs of brown pants needed to be made. The current pace was 161 garments per hour. This was a relatively simple garment for a UK-based mass brand, but just think about that for a second. Stop and take a look at your own clothing. Just the sewing alone for those pants takes 132 hands. Every little bend, belt loop, and reinforced seam is sewn by a real person.

Here, and in many other factories, this line workflow is based on the century-old assembly line system developed by Frederick Winslow Taylor, the father of scientific management and a leader of the Efficiency Movement. Each piece of the garment is assembled by a separate machine and worker for maximum efficiency. Walking methodically along the floor, with clipboard in hand, was a young woman who was introduced as the industrial engineer for the floor. Industrial engineers break down the creation of a garment into standardized timed steps. Each worker's incremental movements are timed to the second to maximize "efficiency" of the workers. A factory would not want one step to take even a second longer than another. Before long, a tiny lag would result in a pile-up of partially made garments, kind of like the traffic jams clogging up the city outside the factory walls.

While making the trousers on order that day, for instance, one worker would sew together the two pieces of fabric for one leg, then clip the partially assembled garment on a swing and push it forward to the station in front of her. The swing runs the length of the row along a metal track. From there, the next sewer grabs the pant leg, folds and attaches the waistband, swing; attaches other leg, swing; pockets, swing; belt loops, swing; interior button tab, swing; zipper fly, swing; bottom hem, swing; enforce seam, swing; iron seam, swing; quality control, swing; and so on. By the end of the sixty-six-person line, the garment looks exactly how we buy it: folded, ironed, label sewn in place. From there the pants are brought in trollies to the packaging room, where each garment gets pierced with a swing tag, logged, and wrapped up inside an individual plastic bag, then piled up with the rest of the order to be boxed and labeled for shipment.

In a factory I visited in China, the managers explained that just one button-down shirt was made of forty separate cut pieces, the assembly of which comprised fifty steps on sixty machines operated by thirty different people. To keep things running, sometimes one worker had more than one task using more than one machine. After adopting a new industrial engineering method, the time it took to make said shirt went from an average of twenty minutes to sixteen—a 20 percent increase in efficiency. Thanks to these "improvements," each worker became as close to a machine as is humanly possible, executing the exact same task with great intensity and precision more than one thousand times a day. More senior workers get the more complicated tasks, such as attaching a waistband or collar, while more junior ones take the more straightforward tasks, like sewing one straight seam. Before me, then, was peak efficiency, a feat of human "industrial" engineering. But by the looks of things, none of these humans had signed up to be engineered.

The women's grimaces contrasted with their joyful attire—traditional saris in every color of the rainbow, with delicate embellishments and sometimes even a matching face mask to protect them from the tiny particles of fabric fuzz flying everywhere. Some women worked the pedals on the sewing machines barefoot, which made me think back to that code of conduct and what safety measures it included . . . sure, they weren't working the saw, but *barefoot*? I wondered how that would go over with OSHA in an American factory, cultural norms aside.

In case anyone was too focused on her work to remember the target numbers on the whiteboard, or didn't feel enough like a machine, the factory had a stark reminder system. In the center of the factory floor was a row of lettered lights, one for each of the twenty-two production lines. If one line was behind target, everyone would know from the glow of the police-car-like revolving red light. Why might a line drop pace, you wonder? Maybe a worker pauses for a few seconds to sneeze, or scratch an itch, or stretch a thumb that had cramped from being contorted into the shape needed to sew belt loops for hours; or perhaps a new worker hasn't yet quite transformed herself into a human machine.

Leaving behind their makers, the garments (and my guides and I) made our way out of production just as soon as we reached the end of the line. Remember, completed garments come out the end of the production line at a rate faster than one per minute, so our forty-five-minute walk-through was already downright leisurely. On our way to the next stop, the packing room, where I would see those pressed garments tagged, packaged, and boxed for shipping, I took a look over my shoulder at the sea of women. Their tense faces, their tense backs, their silence amid the din of the sewing machines and occasional flashing lights, their duller-seeming saris made for an image of consumerism far from the glossy ads we are fed in the West.

Ignorant of the historical echoes of the workers who started our jeans journey—the people of the Americas who were enslaved who made it possible for cotton to be crowned king resonating in my mind—the factory managers told me cheerily how proud indeed they were of their efficiency, citing the investment in industrial engineering as a reason they can produce prodigious amounts of clothing—in case you forgot, that's more than 1 million units per month across their two facilities. But they were also quite proud of the doctor's office and lactation room, which they made sure to show off on the tour. The doctor's office was up a flight of stairs and had a minimal setup—three beds divided by thin curtains; two people were waiting to be seen by the doctor. The space was tiny compared with the vastness of both the cutting and sewing floors. The idea of an in-house doctor sounds very progressive indeed, except when one does the math and sees that the patient-to-doctor ratio was 3,600:1. Similarly dispiriting math revealed itself in the lactation room, which I was awkwardly escorted into next. Despite the thousands of childbearing-age women I saw on the floor, there were only three women in the room breast-feeding.

At the end of the tour, around noon, I was brought back to the CEO's large office for an audience with the factory managers. My hosts, not to mention the garment workers themselves, were all fasting for Ramadan, a month-long Muslim observance of prayer and reflection, but I was politely offered a sandwich and a bottle of water while I waited for the managers. After I ate, the three executives entered the room. We all sat on enormous, thronelike leather chairs, which put us at an awkward distance from one another that matched the distant atmosphere in the room. Most of the executives were in fact from India, not Bangladesh, reflecting a shortage of native-born factory management. I was later told by Shahidur Rahman, a professor at BRAC University, Bangladesh's leading university, that factory jobs in Bangladesh even at the executive level are not seen as prestigious, echoing social stigmas in the United States and China. In order to make real products for demanding customers and protect slim, cutthroat margins, managers need to be available at just about all hours of the day and night. That's also why the executives themselves told me they would not want their own children in the business.

Although the main goal of the meeting was for them to explain their operations to me, the executives were eager to hear about what I saw during my trip to China and how their factory measured up. Even with impressions of the workers' faces fresh in my memory, it was all too easy to reduce them to their output numbers. We slipped into the language of efficiency that people in business are fluent in no matter their home country, and that indeed translates this group's outputs (and the country's) to "impressive."

Any praise we give to factories, in China or Bangladesh or anywhere, for efficiency in this sense normalizes and legitimizes the human-to-machine transformation I witnessed just moments earlier on the factory floor. The dialect of efficiency requires evaluating workers on the same level as the machines they operate: how much can be done per second, how to prevent breakdowns in the system, how to maximize output while minimizing inputs like pay, bathroom breaks, health services, and clear communication about rights.

Throughout my trip to Bangladesh, government officials and development experts spoke fearfully about the ominous, impending arrival of automation, whereby machines would actually take the place of their not-machinelike-enough human counterparts. The way they put it, it sounded frankly apocalyptic—and yet the threat was far more insidious than one day waking up to a world where robots took over the industry. Rather, it was the work of that industrial engineer I saw buzzing along the floor, tracking how long each step took so as to design even more efficient lines, and make machines better able to replicate the fine and nimble work of human hands and brains. Grinning at the thought of a higher margin from higher-output-at-less-cost automated lines, the head of production said that in five years' time, he expected a 20 to 30 percent reduction of employees per line due to automation. Automation might be bad for Bangladeshi development, but it would be excellent for the factories' bottom lines.

My own work—whether it's writing this sentence, reading a report, or sending emails to partners for the New Standard Institute—involves very little of my hands beyond my ability to tap on a keyboard or perhaps expressively gesticulate in a presentation. I could not put myself in those workers' proverbial hands or shoeless feet, which the managers clearly viewed as utterly dispensable; I could not put myself in their minds, and think the thoughts behind their laser-focused eyes as they repeated the same movements over and over and over again without pause. I could not put myself in the position of living life as a human machine.

"MY FACTORY IS A CAGE": LIFE AS A GARMENT WORKER

A few days later, I found myself in yet another subsection of Bangladeshi life: the slum. Just across the lake from my hotel, which was imbued with an air of order, intentionality, and modernity, was Korail, a major slum locals say developed in the 1970s as a residential area after it was acquired by the national telecom carrier a decade earlier. A modern-day, if not slightly apocalyptic, company town, if you will. Entering through a labyrinth of dirt roads, I again felt as if I'd crossed into another universe. The streets were too narrow for cars to pass through but they were nevertheless jam-packed with stalls selling limes and ginger, mangoes and potatoes, stalls selling clothes and other household needs, with barbershops, and with people. Although we were smack-dab in the center of Dhaka, we felt isolated. A trash-filled lake along the back side of the slum separated it from the rest of the city. Fires had ravaged this area in the past, and more fires would burn down a nearby slum just months later and leave ten thousand people, mostly garment workers, homeless.

My guide for this part of my trip—a professor who would be my interpreter—helped me navigate the main area where I was noticeably the only white person. I had asked my American diplomat friend if he wished to join me, but he politely declined, citing security concerns. We walked for about ten minutes, then turned down an even smaller alley, a neat and much quieter residential area. Here was where I'd find the answer to my question: what it was like to be a human machine.

Smiling in the doorway of her family's home was Rima, a thirty-six-year-old garment worker I had been brought to interview through the university. (Why didn't I speak to the garment workers at the factory I visited? Because I was guided by management and knew full well that I would not be given an opportunity to have an open conversation. Asking a worker to reveal the details of their daily life on-site would also put them in a highly compromised, possibly dangerous, situation.) Rima welcomed us inside her home—a single room, maybe ten-by-ten, made of corrugated metal painted bright turquoise—with a sweet, demure expression and a generous demeanor. It was one of twelve on the slum "block," each of which houses two to four people. The whole block shares one kitchen, one toilet, and one shower.

Without hesitation, she offered me the only place to sit in the impeccably tidy, but astoundingly small, space—her single bed, which took up half the room, and obviously doubled as seating for guests like us. Looking around the room, I recalled standing in my New York bathroom, which was about the same size—too small even by New York standards to include a bathtub, but here the entire living space of four people, Rima plus her husband and two children.

Rivaling the advice of any HGTV organization expert, Rima used every inch of available space for storage without a hint of disorder (also in part because there were just so few belongings to begin with). Directly above the bed hung towels and a plastic caddy filled with personal items; immediately adjacent to the bed was a set of drawers, which reached to the other side of the room. Taking up the next wall was a small cabinet with a TV perched on top. At the foot of the bed was a refrigerator, which I learned was a recent and very exciting addition, allowing the family to preserve prepared food. With even this limited decor, there was only floor space for one person to stand—with legs together and arms pinned to their sides, mind you.

Looking around, I wondered how Rima was making this work. How did she and her husband have privacy? Where did they eat? Where did the children study or play? Where did they get dressed for work? How could they stand each other in such close proximity? How could any of them sleep within the corrugated metal walls, especially on the not-infrequent rainy nights? What was obvious, though, was how a place like this could be twisted into a compelling UNICEF fund-raising video, the kind that as a matter of fact does often feature the slums of Bangladesh: *Meet the nice lady who's making the best of having to live in a filthy slum. Just fifty cents a day will feed her and her children. . . .* I've never liked those videos. While they do raise much-needed funds, they also flatten full people into a single label: victim.

Raised in Barishal, a district located in south central Bangladesh, Rima moved to Dhaka about twenty years ago after her family lost their farmland from flooding. Rima has been working in garment factories off and on for years. When she first had her children, a daughter, nineteen, and a son, twelve, she had to stop working in order to take care of them. No one else could take care of her young kids and she told me she would have been too afraid anyway, given the frequent fires in the overly cramped quarters. (This seems common and also helps explain why there were so few women in the factory's lactation room.) She's been at her most recent job for eight years, where she is a sewing machine operator making men's and women's shirts. She started in the industry as a helper in another factory, where an operator taught her to sew. Thanks to this skill, she was able to start as an operator when she moved to this factory, which has about 1,400 employees between their Dhaka and Gazipur locations. Rima works in the Dhaka factory, where 500 workers are split on two lines; the older factories still located in the city tend to be smaller, like this one.

As we began our talk, Rima described her day with a surprisingly steady and bright voice, a tone that belied our surroundings and the details she'd share. She wakes up at five a.m. to make rice and vegetables—

breakfast and lunch for her family. Then she walks forty-five-minutes to the factory, where she produces shirts nonstop from eight a.m. to five p.m. with just a one hour break, which is for lunch at one p.m. On over half of the days she also works overtime, which extends her factory hours to eight or ten p.m. Despite the extra hours, she gets only one additional ten-minute break on those days, during which she is given a piece of bread and a banana. After the nine-to-fourteen-hour workday she walks back home and is expected to prepare dinner. Then she goes to bed, wakes up the next day, and repeats it all—six days a week.

All told, Rima works about sixty-two hours per week, which is about average for garment workers in Bangladesh, according to a recent survey. At 8,000 taka a month—less than $100—Rima's total income is just barely the minimum monthly wage in Bangladesh, even with the maximum overtime pay. But for Rima and many other workers in Bangladesh and Sri Lanka, overtime is the only way that they can come close to earning the legal minimum salary despite the very long hours it requires. And don't forget the unpaid domestic work she is still expected to do for her family—that's another conversation, or book, entirely. Rima is thus among the 64 percent of garment workers who did not receive legal minimum *hourly* wages in Bangladesh, according to a recent survey.

What's worse, a report from the Clean Clothes Campaign (CCC) found that the government-set minimum wage is less than half of what's considered a living wage in most Asian countries. For all of that work, Rima and her husband can barely afford basic expenses: rent, food, and their children's school fees. And she is not alone. In Bangladesh one study found malnutrition among garment workers to be rampant. A quarter of the garment workers were underweight, and of garment workers who are women, 77 percent were anemic.

Rima lives for her children. When I asked her what her hopes were for her own future, her face lit up and she said unflinchingly, "I have no dreams for me, my dreams are for my daughter and son. I want them to be good people and that one day they can take care of me and my hus-band." In those dreams, her daughter will become a nurse and her son an engineer, in contrast to her own middle-school education.

Later, I asked my interpreter if and how Rima's dreams could come true. Maybe, he said with a strained expression. The school fees for her daughter's theoretical nursing school are not that high, and Rima could take a loan. But he was even less hopeful for Rima's son. With their combined salary, he said, "I would doubt very much if they could even get a loan to pay those fees."

I asked Rima if she could tell me what it was like to work as a sewer. As she prepared to answer, the whole vibe of the room shifted. I felt that tension of the sewing floor brewing right there inside her tin house, as if a cloud came by to cast a shadow over the turquoise walls and Rima's orange and blue sari. "Work is intense," she began, her eyes downcast and voice even softer. Her line produces about one thousand units per day. If she falls behind on production targets, her manager appears at her elbow pressuring her to work faster and threatening not to give her overtime. One study that surveyed women from three different slums in Dhaka who work in the garment industry found 67 percent reported they were physically assaulted at work. Rima can go to the bathroom with permis-sion and as long as she stays on target; as one industry expert explained to me, bathroom policies differ with every factory, but workers may be limited to a certain number of breaks per day, and a certain number per line, and need to have their place filled in by a manager. The factory doors are locked—which is illegal, though Rima didn't seem to know this—and there is no official workers' representation. If the workers do protest anything—a seemingly rare occurrence, though it does happen—they make sure the women plan it because they are less likely to be caught by police than the men. One example of a significant worker-led protest was a wildcat strike in late 2018 and early 2019 that led to the death of one worker in confrontations with the police, and nearly eight thousand were dismissed without cause for participation in the work stoppage.

I then asked Rima about the thoughts she had while she executed those machinelike maneuvers at her sewing machine for hours and hours. She looked at me with a confused expression—what did I mean, what did she think about? With my limited-by-comparison experience of rote work, and having read a bunch of papers and memoirs about what low-paying jobs like this can be like, I assumed that Rima's mind would be bored or silently screaming in frustration, that she would fantasize about other things or places as a mental escape. But Rima enlightened me: the only thing on her mind for all those hours was the pressure of production. There was no time for daydreaming, no time to let her mind wander off her shirt. The litany of *no mistakes, keep going, no mistakes, keep going* is about all that there's space for between stitches.

I then asked her what she thought of her job. "When I wake up in the morning, the factory feels like a cage," she replied. Most of the time she feels depressed and unhappy, but sometimes, she added, she feels happy. She has some friends at the factory who live nearby, the mention of whom changed her expression to what might have been a smile.

Rima's story is not unique. This environment of intensely stressful cut-and-sew work plus low pay that she has little choice but to enter into is now endemic in the fashion industry, and is in fact baked into the models at the core of our current economic and labor crisis. As historian Sven Beckert explains, mechanized labor for the purpose of financial profit was at the core of the slavery system of the American South, since there was no other way to ensure productivity than completely controlling the workforce. He cites management scholar Rob Cooke, who argues, "There is no real question nowadays . . . that it [the plantation] was a site of early development of industrial discipline."

When I spoke with workers in Colombo, the commercial capital of Sri Lanka, who had had experience at both larger and smaller factories making garments for brands that, ironically, I and most others in the industry had held up as the most ethical, the daily drudgery sounded, well, gut-wrenching. Most Sri Lankan garment workers are young women in their twenties, who flock to the capital city in search of work. Migratory labor is common in garment work, as the most vulnerable accept the lowest pay. During their stint in the factories, they acquire the derogatory slur "Juki Girls," and are judged by society for their remove both from family life and from a social life, which their long work hours preclude. Living in small group homes without electric light and with constant lines for the bathroom, they are transported to the factory in crowded buses, where sexual harassment and theft of their meager pocket money is common.

Danu, a petite woman whose frame masked immense fortitude, told me that she was forced to shift to day labor to be available for her parents, who live in the north, in cases of emergency. There is no such thing as flexible work in Colombo. Day laborers congregate around the gated entrance to the export processing zone (EPZ), about forty minutes from the center of Colombo, where many of the garment factories, and thus many of the garment workers' rooms, are located. Day labor agencies corral the women into vans; they are told en route where they will be working that day.

I asked Danu, "So this arrangement provides you the flexibility you need, right?" At that point, her eyes turned downward and there was quite a bit of chatter between her and the interpreter. The interpreter looked up at me with an ashen face. The women who enter the truck are sometimes told on the way that there would be no factory work that day, she relayed. Instead they would be going to work at a "massage parlor." I swallowed, buying time before I would have to ask the inevitable follow-up. These women were forced into sex trafficking by the low-paying, precarious work of the garment industry. Though statistics aren't available, case studies suggest that this is not uncommon.

Day labor, the doorway into these heinous crimes, is becoming a common phenomenon in Sri Lankan garment factories, as reported to me by several sources. It's not just in Sri Lanka; women who work in the garment industry in Cambodia have also had to work in the sex industry. There women have been found to be "rescued" from the sex trade only to be forced into garment factories, which, it has been reported, they were trying to run away from to begin with. That the garment and sex industries have a symbiotic relationship is really no surprise, since these offer the lowest-paid work and are done mostly by women. Without other skills to give to the patriarchal society, it's the horrific reality that the only thing left they have to sell is their bodies.

It was excruciating to see the shame in Danu's eyes for the rest of the interview. I felt completely unable to find the right thing, or anything, to say. I told the interpreter that I was so sorry, and asked her to please help me fill in the words. I thanked them both for their time, and just told myself, "Maxine, you will not forget Danu and the situation she represents. You will go back to New York and though you will get sucked back into your daily life, you will not forget about her." I am trying to keep that promise to myself.

I thought of Danu as I sat with Rima in Korail, who had just told me about waking up to feeling like her life was a cage. Again, all I could do in the moment was thank Rima for her time and hospitality, and promise myself again not to forget that slum, that room, that face, and how I was connected to all of it by the clothes on my back.

The professor and I left the room and wound our way back through the narrow alleys to the back side of the slum abutting the lake. I looked out over the vista again to my hotel that was on the opposite bank, which fit into a handsome skyline alongside the fancy wood-and-glass apartment buildings. Between us was a floodplain of trash. I wondered how many people on the other side of the lake even knew Korail existed, let alone about the people who were eking out an existence here while their labor fed the national economy. If I wasn't chin-deep in researching the life of our clothes, would I have given a second thought to the workers on the other side of the world making my clothes?

What is so deeply frustrating about this situation is that it really does not have to be this way. We can do so much better than argue that these horrible conditions are just the cost of economic growth (and hey, at

least these people are no longer starving, they say). Most garment workers in the developing world earn only between .5 percent to 4 percent
of the retail cost of the clothes they produce. That means if you buy a
$20 pair of jeans, only 10 to 80 cents goes to the workers—total. After
churning out 1,600 garments per day that each require fifty different
steps and sixty machines, the thirty people involved in making that pair
of jeans each earn less than a penny per garment.

If Rima's home was not a thin metal box the size of a bathroom,
would that make all of our clothing ridiculously expensive? The research
gives a resounding no. If H&M raised the cost for a T-shirt by 12 to 25
cents, it could allow the worker to earn a living wage. And if brands ab-
sorbed this entire cost of paying living wages within their supply chains,
for mere pennies per product, it would cost them, individually, less than
1 percent of the price of a garment—that's 17 cents on a $25 shirt. That
would create zero suffering for the people at the top of the fashion food
chain, like the executives and shareholders of H&M. The chairman is
the founder's son; he is worth more than $17 billion. His three children,
one of whom is the CEO, and his sister are all also billionaires thanks to
the riches of the fast-fashion monolith. It would affect even less the
founder of Inditex, the parent company of Zara. Amancio Ortega has a
net worth of around $64.9 billion at the time of writing. He and his
family try to stay out of the headlines but couldn't help it when he briefly
beat out Bill Gates to become the wealthiest person on the planet. The
success of these industrious individuals can and should be celebrated—
but how can we value and respect wealth that has cost workers their fundamental basic needs, their very humanity? To the people buying clothes,
the change would also be, well, just change. Would you pay 12 to 25 cents
more for a garment if it meant giving someone else the chance to thrive,
rather than feel like they live in a cage?

On the way back to my hotel, a trip that nearly doubled in length
thanks to traffic, I couldn't stop thinking that Rima and I were the
exact same age. We were both mothers, who work to make our lives and

our families' lives better. Why is it okay for her to be treated in a way that my government has outlawed within its own borders for over a century?

STATE OF THE UNION: WORKERS' RIGHTS IN A GLOBALIZED WORLD

The circumstances of Rima's life and those of her fellow factory workers around the world are indeed an invitation for us to review our own history with some more detail. That's because they're the product of decades of economic, social, and political policies that began, in part, with this history and have since wrapped around the planet like China's (and now Bangladesh's, too) denim production. So now that we have seen the effects of these global changes on the ground, we can zoom out to understand the root cause of the problem, as well as the roots of the solution. And for this we will have a brief history lesson on labor, workers' rights, and international law.

The first and most critical thing to understand in today's globalized world is that there are no strictly enforced universal laws on wages, worker safety, or treatment of the environment. Laws just by themselves are fairly meaningless—they are mere words. That is, their ability to get to the desired behavior exists in proportion to the extent to which they are enforced. Some things that we may think of as laws in the international context—ideas that relate to fair wages and protecting the environment, like the Universal Declaration of Human Rights or the Paris climate agreement—are really just "agreements" that countries sign on to. Countries may sign on to these declarations to protect a belief or principle, but it's all on the honor system to see that those beliefs are respected. To explain this principle, let's use an example all of us know as sure as death: taxes. Imagine that instead of the IRS threatening fines and jail time for not paying taxes, we just had a common agreement with our government that taxes were a necessary part of living in society, so we should all just pay and that's that. Even if we agreed with the general idea, do you think we'd all be as good at filing our taxes? I'll speak for myself: I think not.

The United Nations, which is what many of us think of as the originator of international laws, does not actually have broad, higher policing powers. Its members create resolutions and declarations that are helpful for governments to organize around, but the UN generally does not have the ability to take anyone to court. International law *that is complied with,* which is the only way it can be effective, would require both a rule and mechanisms to enforce the rule, including a police and court system. We have neither at the international level when it comes to labor and the environment. This is why we can't rely on universal declarations to really protect people or places—and begins to explain why fashion, by nature of its globalization and fragmentation, is so rampant with environmental and human rights offenses. Brands are rollin' through countries with little enforced domestic regulation and even less enforceable international regulation like it's the Wild Wild West, but Will Smith and Kevin Kline have hung up their cowboy hats in retirement.

The good news is that we don't need international laws, international police, or even action movie stars to effectively regulate the behavior of other countries. We could use our own *domestic* law to regulate what's awry in the garment industry by putting restrictions on imported products. In other words, America (or any country with a significant enough market) could just say to a brand trying to import its products, *If you want to sell to our big consumer market, you have to play by our rules.* Since the United States is such a large market, a regulation like this would be a significant threat.

The United States does actually have laws on the books, dating back to 1930 and which were reinforced in 2016, that prohibit the importation of products made with modern-day slavery, including child slavery. Yet the US accepts more than $400 billion in goods that were probably made by this kind of labor. And a lot of this is in fashion. According to one report, clothing is at risk for directing the second-highest amount of money toward modern slavery, after tech. The Thomson Reuters Foundation reports that the United States seized only $6.3 million of those imports. Among them: diamonds from Zimbabwe, clothing from China, gold from the Democratic Republic of the Congo, Malaysian rubber gloves, and bone black from Brazil. The poorly funded and understaffed agency responsible for inspections, the US Customs and Border Protection, seized only a tiny fraction of likely slave-made goods. Senator Chris Coons of Delaware, who sits on the Senate Appropriations Committee, told Reuters that they needed additional investments if they wanted to prioritize antislavery regulations, including increasing the $2 million budget for the forced labor team. Currently, that team consists of just six employees out of an entire 62,450-member staff. Without robust enforcement, there are no real incentives for companies, like fashion brands, to ensure compliance.

Europe offers an example of how we might put domestic law into practice with more success. France's 2017 Corporate Duty of Vigilance law sets forth regulations for human rights and environmental abuses for France's largest companies (based on number of employees), the law requires that these organizations must establish a "vigilance plan" to identify and prevent "severe violations of human rights and fundamental freedoms, serious bodily injury or environmental damage or health risks resulting directly or indirectly from the operations of the company and of the companies it controls." Companies must be able to identify risks for violations, have protocols in place for assessing and remedying violations, work with labor unions, and have a monitoring system. Any stakeholder can file a complaint against companies in violation. If deemed guilty, they have three months to rectify the situation. Originally, the law could fine companies between 10 and 30 million euros; but fining has since been removed, so penalties are unclear. That said, companies can still be held responsible for paying compensation to wronged workers. This law itself does not ensure results, but it does provide legal pathways for people and organizations, like the New Standard Institute, to demand accountability. If more nations adopted laws like this, there would be fewer opportunities for the largest offenders to mistreat resources, both human and environmental.

We might think that safe working conditions, or just nonslave labor, are universal inalienable "human rights," but there is nothing divine or absolute about them; they must be codified standards that are meaningfully enforced to be real. And the kicker is this: An origin of the enforced labor standards was right here, in the United States, and women garment workers played a leading role.

FASHION AND THE AMERICAN LABOR MOVEMENT

Close your eyes, click your Made in China shoes, and visualize a black-and-white world you've perhaps seen in history textbooks, one of the biggest sea changes in human history: the Industrial Revolution. Remember how hard it is to make cotton into clothes? And how the whole process of growing, picking, ginning, spinning, dyeing, weaving, cutting, and sewing takes lots of hours and people to get done? People from the black-and-white days didn't have the fancy technology of Qing Mao to get the job done, which made making clothes a very arduous job. Fueled by the opportunity of keeping up with demand for clothes that didn't make people itchy and sweaty, some tinkerers got to work and dreamed up some new inventions that would speed up the pace of the world and bring about the Industrial Revolution. Alongside the cotton gin, the invention of the spinning jenny, power loom, and commercial sewing machines expedited all stages of clothing production. We went from individuals spinning, hand-weaving, and hand-sewing their own clothes to factory-based clothing production all in one generation.

Using cotton produced by the back-breaking work of people who were enslaved in the South, textile mills rose up in New England; these textiles were then cut and sewn into garments in newly established factories in New York City. Enter also the "department store," a new phenomenon introduced to sell this new concept of ready-to-wear garments that would help make New York the first American epicenter of mass clothing production. By 1900 the value and output of the garment industry was three times that of the city's second-largest industry, sugar refining. Ten years later, 70 percent of women's clothes and 40 percent of ready-made menswear in the United States was produced in New York.

New York's booming workforce largely comprised women and children, predominantly Jews from Eastern Europe who had fled pogroms and faced political and religious persecution in the early nineteenth century. They came with great expectations for the freedom America would grant them, but were largely relegated to cut-and-sew factories. Merchants, often Jewish immigrants themselves from more prosperous parts of Europe, utilized this skilled, but cheap, immigrant labor to experiment with the first factory production of clothing. Women also had the requisite skills of cutting patterns and sewing together fabric, having been responsible for making clothes at home, and children were found to be particularly useful on the factory floor because their small hands made them adept at using the equipment. Working in the garment industry could very well be a family affair, and often a necessary way to make sure food was on the table every day and at least some sort of clothing was on their own bodies.

A somewhat similar demographic gained employment in another factory setting, textile production, in places like Lowell, Massachusetts, which became famous for its "mill girls." At first, young, well-off American women flocked to the mills to declare independence, financial and otherwise, in the same way the women I talked to in Bangladesh and Sri Lanka do today. The similarities don't end there. Just because the process of making cotton garments had become faster thanks to machines

in twentieth-century America (and again, slavery, we cannot forget)
doesn't mean it was better for those working the machines. Factory work
was long, tedious, and dangerous. Emphasis at the time was on output
at just about any cost.

Here's where things start to diverge. The women laborers in the States
didn't take things sitting down. And with less external competition than
Bangladeshi women face when they try to protest, their demands were
actually heard. Some of the earliest and most significant achievements
in labor organizing in this country took place in New York and Lowell,
by textile and garment workers, leading the efforts of the Progressive Era
before women in America even had the right to vote. The result of their
efforts were some of the first workers' associations to successfully procure legislation protecting workers. In 1909, union workers from New
York shirt factories, including the largest, Triangle Shirtwaist Factory,
came together in the biggest women's strike in US history to date, later
referred to as the "Uprising of the 20,000." The women shared a common set of grievances about wages, hours, workplace safety, and sexual
harassment—the very same set of issues facing the women who make
our clothes today.

Mass rallies sprang up across the city, and some connected the issues
of working women with the suffragist cause. While there was distrust
behind closed doors between working-class women (mostly Jewish, Italian, and Irish immigrants) and the more upper-class, educated suffrag-
ists (mostly white, Protestant, and born in America), they were able to
forge a unified banner of feminism. Even wives and daughters of the
most prominent American families became involved in the cause. Anne
Morgan, J. P. Morgan's daughter, sought out Rose Schneiderman, a Jewish immigrant garment worker turned labor organizer from what is
today Poland, to be briefed on the goals of the strike.

While the alliances that were formed by these women would prove to
be critical in the future, they were less effective during that time. Only
one of the companies settled with its employees. The Triangle Shirtwaist
Factory did not. With strike funds exhausted, women ended the protest

and went back to work. It would take a tragedy on an industrial scale to turn the tide on worker's rights in the United States.

That tragedy occurred a year later in a building that still exists today just off Washington Square Park in Manhattan (now part of New York University's campus). A fire erupted on March 25, 1911, in the Triangle Shirtwaist Factory, during which young workers were locked inside and were unable to escape. As the inferno spread, sixty-two girls jumped to their deaths, some holding hands, driven crazy by the heat. Fire hoses and ladders were too short to reach the blaze, the fire nets too weak to catch the falling bodies. One hundred and forty-six workers were killed that day. The youngest was only fourteen years old.

In the wake of this disaster, the city and state governments, which had sided with the factory owners during the uprising, at last agreed to put the workers' demands back on the table. A special state commission was set up to review conditions not just in the clothing industry but across a wide range of industries, including the chemical industry. It was the most significant effort yet undertaken by any state. The commission resulted in legislation that granted workers long-denied rights, such as workplace safety standards, and capped working hours in some factories became law in the state of New York, setting an example for the rest of the country.

The ascension of these working-class voices laid the groundwork for the even more sweeping, but not entirely inclusive, New Deal reforms under Franklin D. Roosevelt, in the wake of the Depression. (The New Deal left out many workers, a discussion we will continue in chapter 5.) As the market collapsed, the protective policies first demanded and secured by workers—specifically garment workers—were expanded into dozens of new government agencies, the most significant steps ever taken by the government to protect and advance workers. Frances Perkins, who played a crucial role in the initial factory review commission and would go on to become the secretary of labor responsible for crafting and implementing the New Deal herself, said that the New Deal began with the Triangle Shirtwaist Factory fire. Not a small win for the women making blouses.

STATE OF THE UNION: THE LOSE-LOSE TRADE-OFF FOR WORKERS' RIGHTS

So why can't we copy-paste the same gains over to Rima in Bangladesh, and all the workers like her in Sri Lanka, China, and elsewhere? The answer is simple, but for a complex reason: It was designed that way. As you'll recall, back in the 1980s brands changed their business model and went gangbusters sending production overseas to cut costs, marketing lower prices for their goods while also quietly amassing their own wealth. When the trade agreements were created that stitched national economies together into the globalized world we live in today, worker and environmental protections were *not* included. The rights American laborers fought for and won did not extend beyond our borders.

In our global fashion economy, we the consuming public were told that we would get access to cheaper products, and the developing world was sold on an opportunity for economic growth. Americans did get cheaper jeans, but workers everywhere lost in the race to the bottom. Some of us lost our jobs that would pay for those jeans. Overseas workers got lots of business, but are now eating food grown with toxic water and breathing in coal-filled air, as they toil away at mechanized jobs that ignore the needs, desires, and dreams of their human bodies and minds. Yes, gains have been made, and there are those who argue that the ruthlessness of garment production is just a stepping-stone to more advanced development, but it is far from clear that this is the case, and there is absolutely no reason that it has to be so ruthless.

———

On my trip to Texas I met Cesar Viramontes, owner of an El Paso wash house. Belying his seventy-three years, Cesar's bright eyes, which sparkle against a complexion as weathered as his matching belt and cowboy boots (and, obviously, his blue jeans), have witnessed the parallel rise in denim fashions across the globe and the rise of El Paso to become the blue jeans capital of the world. They've also seen the fallout when that capital up and left the country and continent. Once one of the industry's biggest denim suppliers, he witnessed firsthand what happened to the supply-demand balance of work and workers' protections in the transition from domestic to international labor.

Cesar climbed the ladder of capitalism and entrepreneurship through the American denim industry in a way that not many garment workers, including his own employees and workers like Rima, can today, not even in the United States. After his family immigrated from Mexico when he was five, he worked for a coin-operated laundromat for seventeen years, eventually buying four of his boss's properties and effectively becoming his competitor. His boss supported his protégé's growth, and over time Cesar was able to underprice his "hero." His vast experience and expertise with laundry machines made him an invaluable resource to emerging denim brands such as Guess, Calvin Klein (which was manufactured during that time by El Paso–based Sun Apparel), and Wrangler.

Cesar's wash house has nothing to do with getting jeans clean; rather, it's where "wash and finish," a two-step process unique to denim, takes place. (It's the process that we saw in Guangdong, in the last chapter.) Jeans arrive in wash houses already cut and sewn into the form of pants, but they aren't yet ready to wear. If denim doesn't go to a wash house, putting on a pair of jeans might make you feel more like Gumby (or a seventeenth-century Italian sailor) than an easygoing American cowboy; in its raw state, denim is crunchy and sandpaper-stiff. Levi's, with Cesar's help, popularized the iconic "stone-wash" look that's literally achieved by washing the garment with pumice stones in huge industrial washing machines, giving them the worn-in look and feel that was once achieved by . . . wearing them. Cesar's wash house has transformed hundreds of thousands of pants once worthy of only cleaning wet ship decks into people's favorite things to wear. His house also took care of all the bells and whistles—meticulously casual fades, rips, and frays that make each new pair feel lived in; plus rivets, whiskers, and more—that allow jeans to go beyond function and become fashion. Have you ever felt a bit of grit in the pockets of your jeans when you buy them? That's pumice stone dust.

For decades, Cesar was at the helm of this remarkable denim-to-jeans transformation. During his boom years, beginning in the 1970s, El Paso was a sought-out market for denim because of the relatively cheap labor compared with other US states. That doesn't mean it was easy. Washing and finishing are both skill- and time-intensive work: I watched a woman etch a whisker pattern below the pocket of jeans with sandpaper, her sanding hand gloved to protect her hot pink manicure; across the floor, another woman with a blond bouffant ran a razor blade over the legs of jeans to create a precise pattern of rips by hand.

Handmade "whiskering."

Thanks to the movement galvanized by those twentieth-century ladies in New York, the women and men in Cesar's shop had legal protections of their work environment and wages, even at a relatively cheap expense to denim companies. When I toured the facilities, the space was clean; the workers wore appropriate safety gear; there was a large Fair

Labor Standards Act poster hanging in the break room (where workers actually took breaks) defining in English and Spanish minimum wages all workers are legally required to receive and what they can do about it if they are not; and the effluents, which were held in a neat array of rectangular pools that almost looked swimmable, were monitored by the city of El Paso. Compare this with what I saw in some parts of China, where toxic sludge spewed out onto factory floors and into rivers; and in Bangladesh, where barefoot workers were threatened with verbal abuse and sewed at breakneck speeds for up to fourteen hours a day, six days a week.

But things aren't perfect in El Paso, even in light of these relatively high standards of business. What I saw on my visit was a very different, and very reduced, operation from where Cesar started out, thanks to this two-way race to the bottom. Things started to go sour back in the 1990s. It was the early days of the North American Free Trade Agreement (NAFTA), an agreement between the United States, Mexico, and Canada that merchant brands had actively lobbied for to lower trade barriers between those countries. NAFTA led to an expansion of overseas production of apparel, but *not* an increase in domestic production—or, as important, the well-being of workers and the environment in either place.

Cesar and his team took a hit from NAFTA. A businessman first and foremost, he aimed to be flexible and at first tried to ride the wave by opening up production in Mexico, but it came at a price. He had to let go many of his US workers, some of whom had been with him for decades. At the same time, widespread corruption among Mexican unions, which, as Cesar explained, are effectively controlled by the government, began causing trouble around 2000. The looser labor laws that had made Mexico attractive for production were becoming complicated by the unions, and he was left with expensive labor on both sides of the border—at least, relative to Chinese labor. Meanwhile, the brands he'd had as big-name clients for years delivered the "it's not personal, it's business" line as soon as his prices exceeded those of the new cheaper kid on the block, China.

Cesar has never had unions in his laundry, explaining that the value of collective bargaining is often overshadowed by the facts of the marketplace; in other words, if wage demands are too high, the factory won't remain cost competitive, which can ultimately lead to closure. He'd also watched the demise of a neighboring El Paso manufacturer, Farah, Inc., which, as he saw it, came about at the hands of its own unions. In 1977, after a series of devastating strikes and even a national boycott orchestrated by the AFL-CIO, a major federation of American trade unions formed out of the 1955 merger of the American Federation of Labor (AFL) and the Congress of Industrial Organizations (CIO), the El Paso company closed, unable to meet the demands of its workers. At its peak, Farah employed 9,500 workers and was the second-largest employer in all of El Paso.

Farah's closure foreshadowed the large-scale extinction of the garment industry Texas is facing today, including in Cesar's shop. From 1995 to 2005, El Paso lost 22,000 manufacturing jobs, and nationwide we lost 5 million manufacturing jobs between 2000 and 2010. When Levi's started to close its factories in 1997, workers received the best severance package in the garment industry's history at the time. Levi's even offered "outplacement" programs to teach people how to make it through the transition, mentally and financially, as well as small business allowances to start up on their own. Back in the 1960s, Cesar was able to succeed through this route; he bought his laundromats from his old boss and became the most wanted man in laundry.

But rags-to-riches-type encouragement won't work today for a number of reasons. The skills in washing and finishing that workers honed under Cesar are of little use in a city where lower labor costs elsewhere means denim has left the building, so even a worker with the gumption to pull herself up by the bootstraps and become a solopreneur would have little chance of succeeding. In a supply-and-demand world, supply alone is not the answer. The story would be different if part of her severance package included skills-based trainings that would make her qualified to work in other industries, like tech, or some other kind of long-term compensation. But it didn't. President Bill Clinton even admitted that NAFTA could have included protections for domestic workers who lost their jobs when production moved overseas, though he doubted it would have been approved by his Republican opponents.

The story doesn't end with individual families losing work, though. Globalization and the loss of jobs for some portions of the US population has ultimately had extraordinary political consequences. Researchers found that greater exposure to losses from globalization in predominantly white counties led to an increasing market share for the Fox News channel, stronger ideological polarization in campaign contributions, a disproportionate rise in the likelihood of electing a Republican to Congress, and a shift toward the Republican candidate in presidential elections. (Though it must also be noted that the majority-minority counties that had losses from globalization became more likely to elect a liberal Democrat.) In short, the loss of jobs from globalization played a role in electing Donald Trump. This does not mean that globalization is bad policy per se, but it does mean that ignoring the fact that jobs are lost has severe unintended consequences.

In light of this shift, Cesar and others around the country have started pursuing a new chapter of America's denim story, one that would allow him to stay afloat as a businessman and, hopefully, give workers the kind of security and income they lost because of these trade agreements. He's begun offering full-package production (cut and sew plus wash and finish) to premium denim brands—that is, ones that charge more, like Citizens of Humanity and FRAME—which helps them absorb the higher cost of domestic labor. It's unclear if this will pan out; first, because the premium brands' expectation of having a tighter turnaround time by working domestically isn't so cut-and-dry; and second, because it doesn't truly disrupt the low-wage game that started this whole situation. Cesar's hope for the future rests on the fact that California, where most premium denim production is located now, passed legislation to reach a $15 minimum wage by 2022, and Los Angeles, the hub of premium denim production, required companies to pay a $15 minimum wage in July 2020. This would make the relatively low wages of Texas competitive again.

THE REAL COST OF FREE(DOM) LABOR IN THE HYPERGLOBALIZED ECONOMY

The loss of jobs in El Paso surely seems like the kind of thing organized labor would cover—if, that is, they had the same voice as they did in the early days of American garment making, which they don't. If you'd listened to anything from the Trump administration, you might have thought organized labor was thriving. But that rhetoric was nothing but fake news; union membership in the private sector is at an all-time low of 6.2 percent, lower even than the percentage of private sector unionized workers when organizing was illegal before 1935. We are far behind the rest of the Western world in this respect, especially Europe, where unions cover many more workers than in the States: 90 percent of wage and salaried workers in Iceland, 34 percent of Italian workers, and 26 percent of Canadians are unionized. We can look back to shareholder primacy for the root cause. Businesses began to lobby for stricter laws limiting the power of labor, while at the same time offshoring many of the jobs once held by union members. While this was going on, the neoliberal message was that globalization would save us all, because the GDP and the stock market were growing. But without a means to collectively bargain for better wages and terms, working-class Americans have been unable to participate in those gains. Wages in the United States stagnated and the middle class shrank, creating the massive rich-poor divide we're mired in today. (We'll look at that more closely in chapter 5.)

For the unions that do exist, there is also a major qualitative difference between unions in America and elsewhere. Cooperation between corporations and the unions is one of those European things Americans just don't get, like vacation from work and sitting down for coffee, which creates a checks-and-balances system in terms of demands. Contrary to Cesar's fears of worker-based collapse, in Europe both parties know how far they can go before being shut down, so there is a much higher rate of success. In Germany, for instance, union members are also on the boards of corporations, so they are able to communicate transparently and with full knowledge between workers and managers.

In the States, though, the endgame of unions is often exactly the opposite of what the name implies: rather than uniting workers in all-trade collectives to standardize wages, hours, and benefits across an industry, unions are often company-based. This means theoretical unionized Levi's washers wouldn't necessarily benefit from any protections won by theoretical unionized Wrangler washers, and vice versa, creating intra-industry competition that hurts workers, but not companies. We see how broken the system is in the results of an Organization for Economic Cooperation and Development (OECD) evaluation, which rates the quality of workers' protections in different countries. The United States was last in strictness in temp work, and had the lowest score out of seventy-one countries in an evaluation of rules for severance pay and reason for dismissal.

It's undeniably positive that unionized American workers have raised their voices for better conditions, but the lack of awareness about global labor pits US workers not only against themselves, but against the rest of the world in a cruel Robin Hood trade. Without international labor standards included in our trade deals or enforced through our own domestic laws, the more "costly" protections workers in the United States demand (whether it's higher wages or benefits like health insurance and retirement plans), the more likely companies will be to seek labor that's less costly, from workers across the border and overseas. Looking at things globally, we're back to where we started in certain respects, facing a deluge of underpaid but overworked laborers with perhaps more similarities than differences to the immigrants that settled here over a century ago.

Dani Rodrik, an economist at Harvard's Kennedy School and a critic of neoliberalism, calls this current setup—the almost complete lack of barriers around trade that as a result lowers welfare standards—"hyperglobalization." In a hyperglobalized world, multinational corporations, including clothing brands, have taken control from governments and are setting the rules for the global economy. As multinational brands like Gap, H&M, J.Crew, or Amazon shop around the globe for the lowest prices, they pit countries against one another, creating massive disincentives for local governments to create and enforce laws, and by extension develop their own domestic labor or environmental protections like the EPA or OSHA.

Fashion has taken the lead on the race to the bottom compared with other industrial sectors because, unlike what you need to make other stuff like iPhones, it is not terribly challenging or expensive to set up a garment factory. All you really need is sewing machines—which are not costly—and people behind those sewing machines—which we have found are also not costly. Since factories can be set up virtually anywhere, the cheapest price can also be found anywhere. It should thus come as no surprise that the major fashion production centers today also have the weakest democratic norms. Transparency International, a global organization devoted to fighting corruption, released some alarming stats for 2018: On its Corruption Perceptions Index, which ranks the public sector corruption of 180 countries, of the major clothing-producing countries China came in 87th place, Sri Lanka 89th, Vietnam 117th, and Bangladesh 149th. The race to the bottom indeed.

These disincentives for enforced standards and regulations were on shockingly full display in April 2013, in Bangladesh on the outskirts of Dhaka District. Rana Plaza, a large factory that served many Western brands as customers, collapsed under the weight and vibrations of its own machines, which were in violation of local, and clearly weakly enforced, building ordinances. Inside the locked fire doors (also technically against regulation), workers were trapped. Workers had complained about cracks in the building just a day before, but when they protested they were threatened with docked pay and told it was safe. The next morning, 1,134 people died and 2,500 more were injured in the collapse. The fashion industry, and the system it created to inhibit worker protections that would have saved those people, is responsible for the worst industrial accident in the history of the world to date.

I'm not someone who defaults to the fatalistic view of history repeating itself—if I were, I wouldn't be writing this book. But the near mirror image of the disasters that took place at the Triangle Shirtwaist Factory and Rana Plaza, 102 years and one month apart, is startling. We can rewrite the rest of this narrative if we choose. To get there, though, we'll need to peel back another layer of how our clothes are made, one you probably didn't even know was there.

Four

Middlemen, Management, Marketing, and a New Kind of Transparency

I f you were buying a pair of Levi's jeans in 1960 they would likely have been made in the USA. The CEO would know where Levi's factories were located around the country because it owned most of them. In the age of hyperglobalization, though, if you buy a pair of jeans at, say, Gap that says "Made in China," you would have no idea if the fabric, zipper, or anything else came from other places, just that it was cut and sewn together in China. The CEO of that clothing brand may not know those details either, because she's thinking about the marketing budget for Christmas and just wants to know that the new style of jeans that's hot on Instagram will deliver on time.

The reason our modern-day CEOs may be so lax is because of the next player involved in our garments' journey, one you're less likely to know about than any other stage in this book. That's on purpose. As more and more countries and brands tried to get in on the game of globalization, some entrepreneurial spirits saw an opportunity. Brands needed help navigating the Wild West of suppliers, so a new entity—the apparel broker firm—was born. Li & Fung is the largest of these companies; it runs a worldwide network that orchestrates the production of an enormous amount of our clothing in forty different markets. Founded in 1906 by an English teacher and a merchant dealing in silk and ivory, it now has 230 offices around the world. In 2018, its sales surpassed $12 billion. In the words of the CEO back in 1998 (hence the reference to Multifiber Arrangement quotas below), here's how the group would typically complete an order for, say, ten thousand garments:

> We might decide to buy yarn from a Korean producer but have it woven and dyed in Taiwan. So we pick the yarn and ship it to Taiwan. The Japanese have the best zippers and buttons, but they manufacture them mostly in China. Okay, so we go to YKK, a big Japanese zipper manufacturer, and we order the right zippers from their Chinese plants. Then we determine that, because of quotas and labor conditions, the best place to make the garments is Thailand. So we ship everything there. And because the customer needs quick delivery, we may divide the order across five factories in Thailand. Effectively, we are customizing the value chain to best meet the customer's needs.
>
> Five weeks after we have received the order, 10,000 garments arrive on the shelves in Europe, all looking like they came from one factory, with colors, for example, perfectly matched. Just think about the logistics and the coordination.

Like brands' personal shoppers, Li & Fung has been responsible for "optimizing each step in production" for companies such as Walmart, Kate Spade, Coach, Calvin Klein, and Tommy Hilfiger, among others. And yet I bet you've never heard of them.

By playing middleman so efficiently, Li & Fung and others like them added layers of opacity to the supply chain for brands and consumers alike. Sourcing companies were not quick to reveal to the brands who was responsible for manufacturing. Likewise, the brands had equally little interest in knowing much about the factories. This system is part of the reason why when I asked vendors about their suppliers on my sourcing expedition for Zady at the Javits Center, I got blank stares.

With neither government regulation nor brand oversight into the conditions of factories on the ground, mini versions of New York City circa 1911 cropped up in apparel-producing countries like a silent plague. An underground world of sweatshops became an unspoken player in the race to the bottom, but they wouldn't stay hidden forever. Eventually, journalists began exposing the worst offenders. There was the Kathie Lee Gifford scandal of 1996, when it was broadcast that her Walmart line was made by thirteen- and fourteen-year-olds working twenty-hour days in factories in Honduras; Nike also had a PR nightmare when investigations into its production in Indonesia were reported in the nineties.

Suddenly, brands couldn't hide behind the invisibility cloak of ignorance Li & Fung and others had so nicely draped for them. As high-profile media threatened their reputations, they had to show they were doing something. Enter the vendor code of conduct, a set of policies crafted by brands stating the expectations they have of their manufacturing partners. The codes cover things like the environment and waste disposal; health, safety, hours, and overtime of workers; as well as the expectation that manufacturers do not hire child labor and slave labor. As a 2019 Clean Clothes Campaign report explains, "these codes have become the central tools through which brands seek to demonstrate to their customers that they are addressing worker rights in their supply chains."

Let's decode (pun intended) this last statement. Brands' codes of conduct are meant to "demonstrate to their customers" that workers' rights are protected, which is a very different goal from ensuring that workers' rights are actually enforced. I repeat, codes of conduct are there *not* to create or protect rights for workers, but to avoid responsibility for unsavory factory outcomes. No wonder the code of conduct posters I saw in the Gazipur stairwell were hardly legible for the people actually working in the factory. They could have been written in jabberwocky—as long as they were up on a wall, the brand could fall back on the fact that they *had* a policy. If brands don't do anything meaningful to ensure that those codes are implemented, then they're just ugly posters.

CODES OF CONDUCT AND CORPORATE SOCIAL RESPONSIBILITY

Let's put ourselves in the shoes of brands, just for a moment, to imagine what they think they're saying with these vague codes and nonenforced policies. In an era in which brands must at least pay lip service to sustainability and ethics, brands almost always proudly publish their company mission statements on their websites. "Radically transparent" start-up Everlane writes, "At Everlane, we want the right choice to be as easy as putting on a great T-shirt. That's why we partner with the best, ethical factories around the world." Merchant brand stalwart Gap: "Good Business Can Change the World." Madewell, the younger, cooler, sister of J.Crew, proudly proclaims: "At Madewell, we strive to do well in the world." J.Crew itself: "We imagine a world where doing good is part of doing business." (J.Crew Group filed for bankruptcy in May 2020, the first major brand to do so in the wake of the COVID-19 pandemic and its tidal effect on garment sales.) And Inditex, the parent company of Zara, tells us, "People are at the heart of everything we do."

All sounds very warm and very fuzzy, doesn't it? Just reading these proclamations makes me want to run out and buy something—not because I really even *want* it, but just to do my part to make the world a better place. Already, we're being hoodwinked into buying under the auspices of improving the world, a core tenet of neoliberalism. But if you look beyond the aspirational mission statement and read the actual policies, you'll find that it's not all rainbows and unicorns.

Everlane, for example, writes that it "aims to conduct business to the highest level of ethical standards and *expects our business partners* to operate in full compliance with all applicable laws, rules, and regulations, as well as to abide by our Social Responsibility Code of Conduct and promote continuous improvement of working conditions throughout our supply chain." [Emphasis mine.] Gap, Madewell, Inditex, and J.Crew make similar pronouncements about what manufacturing partners and suppliers must do.

In their mission statements, every company was out to do good and be the change they want to see in the world. But in their actual policies, there is a subtle but critical bait and switch: figuring out *how* to execute that mission falls not on the brand, but on the manufacturers. So when a factory collapses, or a newspaper reports on a manufacturer's use of slave labor, the brands can wiggle out of the responsibility: We don't actually *make* the clothes, they claim. It's not our fault, it's the factories', they clamor. The space between some of the companies we think of as "ethical" or "sustainable" and those we think of as "fast fashion" collapses. All these brands seem to participate equally in shifting responsibility.

THE AUDITING RACKET

Now, let's take this one step further. Brands seek to demonstrate that these codes are being complied with through audits. Brands hire large, mostly Europe-based international auditing firms to perform the audits by going in and checking that a manufacturing partner complies with their code of conduct. Again, the goal is rarely to ensure that the workers are doing well and there's no funny business going on, but to mitigate risk.

Like fellow middlemen the apparel broker firms, auditing brings in billions of dollars every year and employs thousands of people. Some significant firms include France-based Bureau Veritas, Germany-based TÜV Rheinland, Switzerland-based SGS, and US-based Underwriters Laboratories. The Ethical Trading Initiative (ETI) estimates that the total third-party audit industry is worth around $80 billion, comprising up to a crazy 80 percent of companies' "ethical sourcing" budgets. Demand for ethical clothing has been a boon to these companies; a 2013 *New York Times* article reported that three of the largest publicly traded monitoring companies enjoyed a 50 percent increase in their share prices over the previous two years. In the two decades since auditing firms became popular, thousands of factories have been approved each year. All this to demonstrate how central auditing is for brands as the means to try to demonstrate that they are not the bad guys.

Auditing has become the centerpiece of so-called ethical manufacturing because it allows companies to show that they are doing something so they can market themselves as a company that looks out for the workers and cares about the environment. Having the audits performed by third parties provides a built-in scapegoat. If some tragedy happens, it's clearly the factory's fault and perhaps the auditor's fault, but definitely not the brand's.

The game of responsibility hot potato goes like this. Brands tell their customers they're saving the world, then pass off the world-saving responsibility to their factories and use the auditors to give them clean bills of health (though the auditors are themselves not in the business of ensuring that factories are saving the world, but rather ensuring that the brands are protected from the damning *New York Times* cover story if that hot potato were to get dropped). And, surprise, surprise, a lot of potatoes have dropped, which turns the whole cycle back around—the brands say, "not it," point to their codes, and let everyone else scrape the potato off the floor.

One instance of auditing gone awry came to light when female workers at a factory in Bangladesh, producing for Lululemon, reported being beaten, subjected to verbal harassment (called "sluts" and "whores"), and forced to work when sick. When this was revealed, Lululemon stated, "We require that all vendors share our values and uphold a consistent set of policies that live up to our Code. We do not tolerate any violation of this Code," but no action was announced. When the Brazilian government uncovered a factory producing clothes for Zara where workers were paid below minimum wage and at least one of them was fourteen years old, Inditex, Zara's parent company, brushed it off by claiming that ·this factory was "unauthorized outsourcing"—the parent factory had defied Zara's code of conduct. And when it came to light that seven- and eight-year-old children were working sixty hours a week in ASOS factories instead of going to school, the brand fell back on a familiar refrain.

We can see now how quickly the buck gets passed from brand to manufacturer to no one, which makes the "big brand saves the world" proclamations just another headline of fake news.

Overall Weaknesses in Auditing

Now, as we would say in law school, let's unpack the systemic weaknesses of this system and see exactly how these codes and audits that are meant to ensure their implementation fail to protect either the workers or the environment.

At a logistical level, audits are arduous, time-consuming, and most significantly expensive, and all of these costs are borne, unsurprisingly, not by the brands, but by the factories themselves. Inspection fees range from a few hundred to over one thousand dollars per day, and last anywhere between one and twelve days; this averages out to $2,000 per compliance check. Each brand has its own same-but-ever-so-slightly-different code (though there has been ongoing work to streamline this) and demands its own separate audits, making auditing expenses a significant burden on the factory. Factories, which already operate on really tight margins (and really don't want anything bad to be found, which would cost them business), want to get auditors in and out as fast as possible—sometimes in as little as one day—naturally resulting in less-than-thorough inspections. According to one report of 2015 audits, twelve factories used by Gap, Nike, Target, and Walmart were all inspected in just two days.

Auditors use worker interviews as a significant part of the audits.

Factories know this and have been found to train the workers on how to respond to avoid uncovering any violations. I spoke with one such auditor in Sri Lanka, whose story was honestly one of the most bad-ass I've heard in all my life, even outside of research for this book. A former garment worker herself, Ashila Dandeniya single-handedly stood up for herself in court to demand higher wages, unprecedented for a woman in Sri Lanka. She now runs an organization, Stand Up Movement-Lanka dedicated to educating workers on their rights and providing resources and outlets to share their experiences safely—and she rides a motorbike around town. She described with a note of dismissal in her voice the ridiculous methods auditors use to collect interviews from prescreened workers, who have been trained on how to answer to avoid bad marks. A skilled auditor, which most are not trained to be, would know the difference between, for instance, phrasing questions as yes or no versus qualitatively, which yield different answers; compare how a worker might respond to "Did you eat lunch today?" (a question she might have been trained to just say yes to) versus "What did you eat for lunch today?," the latter requiring something other than a coached answer.

In the case where a factory is flagged for violations, brands are also, not surprisingly, slow to do anything about it. A Sri Lankan factory owner I spoke with cited this as one of the central weaknesses of auditing. If brands want safer factories, he explained, then they need to have actual incentives for factory owners to make corrections to dangerous practices—like ending a relationship with a factory if they continually fail to comply with the codes. He offered Nike as an example of the exception that proves the rule. It is the only company that he knew of that actually drops factories after two failed audits (though perhaps this is because it remembers being the poster child of sweatshop factory conditions in the nineties).

So on the one hand, brands don't act when factories are found to be out of compliance; and then on the other hand, they don't provide much in the way of real incentives to comply with their codes. No stick, no carrot. In fact, the *Better Buying Index Report*, Spring 2018, found that over 60 percent of suppliers were not incentivized at all for being compliant to buyer codes of conduct. Ultimately, decision makers within the brands themselves may not be aware of rule breaking, by design. Corporate social responsibility (CSR) teams, which are responsible for overseeing auditing, and helping craft those messages of awesomeness, generally run on very separate tracks from the sourcing teams, which are responsible for finding factories, placing orders, and making payments. This means that the people on the CSR teams who are supposedly regulating compliance on things like hours and pay have little knowledge of and even less control over what's being ordered, or critically what the brands are demanding as a price. Tellingly, CSR policies are often drafted by the legal or communications department, not the sourcing teams. In the case of Macy's, CSR is run by a senior executive who heads the corporate communications department. At Walmart the person responsible for social responsibility and sustainability is the same person in charge of "government relations," or lobbying, in layman's terms. It's a fairy tale grafted onto communications, designed to protect the company from risk.

Bad Purchasing Practices

Speaking of prices, the way brands set prices with the factories and hold to (or not) terms of payment is yet another oversight of the auditing system. Factories generally stay in business because they provide low prices, meaning low wages for their employees. And while brands brag about how they are helping the world, the same *Better Buying Index Report* found that 43 percent of factories reported experiencing "high pressure negotiating strategies" from brands about lowering their prices. Again, these companies are abiding by the ethos of profit over people at the core of neoliberalism.

We see this playing out in pricing trends. Despite the devastating Rana Plaza factory collapse in 2013, which led to costly upgrades for factories, per-unit prices in Bangladesh actually fell 2.12 percent in the fiscal year 2016–17 compared with the previous year. In the next fiscal year prices dropped another 4.07 percent. When you continue to pay less and less in an industry that already operates on the thinnest of margins, something—usually working conditions, pay, and environmental protections—is going to give. The math just doesn't work out otherwise.

I work in a world of mostly fake deadlines—no matter how urgently needed a deliverable seems, there always seems to be a few more hours (or weeks). Not the case for apparel, where delays create logistical nightmares for factories and ultimately more pressure on the workers. Numerous factory managers told me about the melee that ensues when a customer comes to them saying they actually need more product, or need it sooner, or some detail in the design has to be changed, and that if they don't deliver they will be penalized. According to Human Rights Watch, only about "16 percent of buyers [brands] met all agreed-upon deadlines in the product development and pre-production stages; another 20 percent of buyers failed to meet these deadlines about 80 percent of the time."

Even if products are made on time, whether or not factories get paid isn't guaranteed. According to a 2016 International Labour Organization (ILO) survey, 52 percent of apparel suppliers claimed brands paid prices lower than production costs, the highest proportion occurring in Bangladesh. Late or nonpayment ultimately causes many issues for workers down the line, which are covered up by poor audits. Without the cash flow to pay workers, factories concoct fake books and ledgers, a "widespread practice" according to a 2013 ILO survey. These double books also allow factories to demand excess overtime from workers, especially when facing rushes from an altered order delivery date.

This power imbalance between brands and factories made major news at the start of the COVID-19 pandemic in 2020, when the sudden closure of retail stores around the world radically cut demand for garments that had already been on order or in production. Brands began canceling orders as if the clothes themselves were infected with the virus (some even asked for discounts on orders they agreed to accept), and since most brands only pay suppliers once the product has shipped, the factories were left in dire straits. What would they do with this excess of clothes? How could they hope to stay in business, let alone pay their workers for the work they've already done? (And I'll remind you, the workers were also exposed to the myriad fears, pressures, and illness of the pandemic itself. No paycheck means no food.) Thankfully, media and advocacy pressure resulted in several brands, including H&M, Inditex, Marks & Spencer, Kiabi, PVH, VF Corporation, and Target, committing to taking and paying for all existing orders. While this was an important display of support, it is like putting a Band-Aid on a patient who's just undergone open-heart surgery. The bigger, central problem is the fact that this swindling could happen at all, due to the power imbalance embedded in the lawless relationship.

Brands are able to change, ignore, and leave out the most crucial terms of their business agreements, in normal times and in pandemics, because of the gross power imbalance in apparel brand–supplier relationships that, like the unenforced labor and environmental codes, was baked into the system from the start. Just about every factory I have gone into has raised this same subject unprompted. It all comes down to the age-old supply-and-demand problem. Since brands have seemingly countless options when it comes to factories to make their garments (remember, factories are easy to set up), they have the upper hand. They make insane demands for the lowest prices, best quality, and fastest turnaround times, and factories, working with in an anemically regulated global system, have to do whatever they can to meet them if they want to stay in business. But every time a brand demands a redo on a batch of garments, or a lower price, they further compromise the factory's ability to comply with whatever sustainability and labor codes might be in effect and limit the ability to raise the social or environmental bar.

One factory executive in Sri Lanka with very prominent customers shared the kind of exasperated conversations he'd have with brands about the disconnect between codes of conduct and what brands are actually willing to pay to have them implemented, saying, "You cannot have a conversation around [workers'] pay without a conversation around what you pay us." In other words, brands can't demand that factories comply with the wage requirements (or other social and environmental requirements) in codes of conduct if they aren't willing to pay the factories for the work. "If you speak about fair wages, can you speak about a fair price?"

WORKERS' LOSSES

There are reasons to gripe about all of these failures of the audit system as a stand-in for universal enforced regulation for pretty much everyone involved on the supply side. And yet the biggest burdens are borne by the workers themselves.

The first offense is pay. As we saw, there are minimum wage laws in countries like Sri Lanka and Bangladesh, and brand codes of conduct decree that their suppliers must obey local regulations about minimum wages. Gap, for instance, writes, "Workers shall be paid at least the minimum legal wage or a wage that meets local industry standards, whichever is greater"—verbiage that's typical of other companies, like J.Crew and Everlane, as well. However, there is some evidence that garment workers are paid less than the legal local minimum wages despite the codes' existence—and even those minimum wages aren't enough to live on.

And what exactly is the "local industry standard"? Do you know based on this phrase, or anything you've read so far in this book? I don't, because as the process of writing this book has taught me, there is no such standard, local or otherwise. Without clear parameters around what they're looking for on a factory's payroll, how could an auditor confirm whether a factory is or is not in compliance and paying their workers enough? (That's another legal lesson: When you want to make compliance difficult, just use vague language.) Naturally, the lack of clarity makes for a lack of enforcement, which research backs up with data for systemic, industry-wide violations. According to a 2013 study (the most recent published data we have) conducted by Better Work, wage-related violations were present in all five countries surveyed: Vietnam, Indonesia, Lesotho, Jordan, and Haiti, some of the newer players in the race to the bottom. In 2012, an ILO report found that 51 percent of Indian garment factories were not in compliance with the minimum wage laws. A failure rate of over half is the clearest demonstration that the code of conduct/ auditing regime is, well, bogus. And now we see how Rima, who herself was paid an hourly rate below minimum wage, fits into the larger picture. Her case isn't an outlier—sadly, she's the norm.

The second layer of wage injustice is this: Even if factories paid the legal minimum wage based on enforceable codes, those wages are not a "living wage," which researchers define as enough for a worker to meet their needs and their family's with some left over, without overtime. The Clean Clothes Campaign report found that government-set minimum wages are less than half of what's considered a living wage in most Asian countries. Remember previous discussions about governments being disincentivized to create protections for workers lest brands decide to just go elsewhere? Here it is playing out. If any one government were to stand up to the big companies investing in their economies (hello, neoliberalism) and say, "We can't make your jeans because we can't pay our citizens enough at the price you want," they'd shoot themselves in the foot. So everyone stays quiet about it and hopes the vegetables grown with toxic water keep people alive long enough to finish the next order from Company X.

But it can get even worse. According to the people I talked to in Bangladesh and Sri Lanka, with price and turnaround pressures so high, factories not only cut wages, they also cut costs in more indirect, and again scarcely documented, ways. First, they move workers from full-time employees (which may come with benefits) to short-term contracts, similar to what workers in the West have been experiencing when they become contractors as part of the so-called gig economy. Danu, the Sri Lankan woman who was at the mercy of manpower agencies, is one such contractor. What she endured in being forced to work as a day laborer was devastating.

Factories are also relying on subcontractors, aka "shadow" factories (some of which might be deemed sweatshops—there isn't a lot of data about which ones are, and which aren't), to fulfill orders at the standards demanded. You'll recall from the reports cited earlier that a brand's response to shady things happening at a facility that a factory had to send the work out to (perhaps because they couldn't keep up with the required schedules or pricing) is something like: *Well, we didn't okay that, so it's not our fault workers were underpaid/overworked/beaten/sex trafficked.*

To my disappointment but not my surprise, I couldn't get into any of these unauthorized subcontracted factories on my trips. The actual number of these so-called shadow factories in existence is unknown. Some surveys of the industry in Bangladesh estimate about half of all the country's factories are illegal, and we do see a correlation between subcontractors and wages, ironically. In Cambodia, for example, the number of export-focused subcontractor factories rose from 82 to 244 between 2014 and 2016, at the same time the basic monthly wage increased from $100 to $153 between 2014 and 2017. To put it simply, when the minimum wage increases in these countries, manufacturers find new, illegal ways to exploit workers. Subcontractors earning below the minimum wage keep business flowing even when workers are *supposed* to be making more. It's the only way factories can stay on top of orders' quality and quantity, appear to pay some workers the legal wage some of the time, and meet CSR initiatives.

Having seen the recipe for an industry that hunts for the cheapest prices combined with a toothless CSR regime, we get a final product of an industry that plays a leading role in child labor and modern-day slavery as mentioned in the previous chapter. In 2018, the top three garment-producing countries in the world (China, Bangladesh, Vietnam) were found guilty of forced labor, child labor, or both in their respective garment industries.

THE SOLUTIONS:
REDISTRIBUTING WORKER ADVOCACY
AND LABOR RIGHTS TO ALL

Reading about the problems of the present-day garment industry may
have brought up a sense of déjà vu. American garment workers in the
early twentieth century were being abused and exploited in not dissimilar ways, as the image below shows in bone-chilling clarity. The difference is, when the US apparel industry went through that period of
unsafe conditions, the union movement rose up to meet it, which itself
helped to usher in the Progressive Era and sweeping New Deal benefits—
such as overtime pay and the concept of the weekend.

*A child in an Industrial Revolution–era garment factory (left), and a child
in a current-day garment factory (right). The similarities are eerie.*

There is theoretically no better way or more direct way for workers to
have a voice vis-à-vis the factory owners and, by extension, brands than
unions. History has shown that workers fighting for their own rights is
what lifts up workers, not wishy-washy codes of conduct that rely on the

honor system and self-regulation. But there's a catch. As we heard from Cesar in Texas, who feared being put out of business by a union upheaval like his competitor, union efforts can backfire when their demands make a factory uncompetitive globally. Unfortunately, this happens often in the fashion industry, where brands have long been driving wages and conditions down with unreasonably low prices for factories, and the barriers to entry are so low. Imagine if a brand is considering working with Factory A in one country, where workers are unionized and paid $10 an hour, or Factory B, where workers are not unionized and wages are $5 an hour. Which one do you think they'd choose? It is why regulators have to step in, either through enforced trade agreements and/or domestic law that extends worker protections—including the right to organize—that would be consistent from country to country to prevent the race to the bottom.

Brands' codes of conduct generally include the right to form a union and collectively bargain, but as we've seen, the terms of those codes are more or less meaningless when they cannot or are not followed for the myriad reasons they fall through the cracks—auditors ill equipped to know what they're looking for or ask the right questions of workers and management, factories trying to stay afloat and shirking the rules at the expense of their workers, and brands hiding behind codes to avoid media scandals. Moreover, the codes aren't targeted to the particularities of the locations where they're meant to be enforced—consider the undefined minimum wage stipulations, or the fact that truly independent unions are effectively illegal in China, so to suggest the idea of collective bargaining there is just about a moot point. When brands' CSR whizzes write nice-sounding copy into a code of conduct without taking into account the broader global system, they perpetuate the race to the bottom, which among other things creates powerful disincentives to allow unionization that would put increased pressure on pay.

For a brand to really show that they are committed to labor having a voice, they have to stay committed to factories and stop shopping around for the best bargain on workers. They can give factories a more equal footing and fill in some of the gaping holes we saw earlier in the chapter. This looks like fairer purchasing practices that take into consideration the demands on the workers and the cost of environmental compliance, as well as ensuring that their auditors are well versed on what union representation means and looks like. As the CCC report on auditing noted, insufficiently trained auditors can report that workers' rights to freedom of association are being met when they're in fact not being met.

If brands working with factories is the ideal scenario for change, looking outside of factories—to industry organizations and NGOs—is the second-best solution. But they should be supportive patches on the industry's holes rather than the only repair. Better Work, a collaboration of the ILO and the International Finance Corporation (IFC), is one group that focuses on improving industry standards in several key manufacturing countries, including Bangladesh. It's implementing tools for workers, unions, management, and government.

However, I found the line between these initiatives and CSR/ marketing-speak to be thin. One major brand I spoke with was eager to report on the "capacity building" programs that Better Work runs in their supplier factories, wherein low-level managers are trained on how to report at-work grievances from workers, and workers are given surveys where they're asked if they know the manager to whom they should ostensibly report a problem. But this grand program missed any assessment of whether this training actually resulted in any quality of life change. Paying for a capacity-building program and throwing that in a CSR report is far from a real fix, as it fails to address the inherent power imbalances at play.

And while it's important to improve these at-work grievances, including how to report unsafe factory conditions—very important based on what we've seen so far—when I spoke to workers the thing that they universally cared most about was wages. No factory workers I spoke to saw Better Work as a draw, which makes me question its efficacy.

Looking at all these pieces together may be overwhelming. Without enforced laws in place, it's hard to imagine a world where brands, competing in an extreme neoliberal global market, would willingly sacrifice profits or choose to abide by an external regulator's standards. But we do have an example of just that. Like a phoenix rising from the ashes, a working version of a non-government-regulated, enforced regulatory system has arisen from the aftermath of Rana Plaza. This unprecedented loss of life resulted in the Accord on Fire and Building Safety in Bangladesh, known colloquially as "the Accord," a program for building safety based on a binding agreement signed between brands and unions that has managed to actually make significant headway in improving factory safety in Bangladesh.

Between bites of oatmeal, the executive director of the Accord, Rob Wayss, filled me in on how they've managed to #getshitdone in such a messed-up system. He told me that while the Accord was officially signed after the collapse in 2013, its principles had been in place for a few years. You see, back in 2010, there was a fire in the Ha-Meem factory, a major supplier of Tommy Hilfiger. During Fashion Week that year, which commenced shortly after the incident, Hilfiger was harangued by media who demanded to know how he would address the situation. Fearing a press and consumer superstorm, he publicly agreed to open the factories that produced for his brand for inspection. The icing on the cake was that ahead of a Diane Sawyer TV news report on Bangladesh, Hilfiger's parent company PVH made another statement of commitment to factory safety. From this, an agreement that is the basis for the Accord was drafted, and PVH agreed to sign, although four others needed to join in order to activate the agreement. So the agreement just sat there.

Then Rana Plaza happened. The media swarmed like flies on you-know-what to cover the largest industrial disaster ever, and people were outraged. This gave the union and NGO community—including IndustriALL (a global labor organization), the Clean Clothes Campaign, and the Worker Rights Consortium—the leverage they needed to rally

major companies H&M, C&A, Carrefour, and M&S to become parties
to the agreement with the unions and finalize the terms. The agreement
was signed in less than a week.

Based in Europe, the Accord has elevated the standards of factory
safety in Bangladesh through regular health and safety inspections of
building codes from a committee of workers, unions, and management.
The Accord further establishes employee safety training and a worker
complaint process. It's not the international regulatory regime Dani Rodrik posits as the end to neoliberal hyperglobalization, but it's a step forward. The organization set up to implement the agreement started out
with a small team of international engineers and locals doing inspections, and grew to a reach of more than 1,600 factories, eight domestic
and two international labor unions, and more than 190 signatories. It
was entirely funded by the brands, which paid a fee to be included based
on each brand's annual volume produced in its factories in Bangladesh
and the US dollar value of its exports. Unlike traditional audits, where
the results are not made public, Rob is particularly proud of the fact that
all of the inspected factories' statuses were listed on the Accord's website.
The Accord was meant to expire after five years of work, and after a
transition agreement was signed that allowed operation until 2021, the
Accord was superseded by the Readymade Sustainability Council in
2019. As of writing, it is not clear how successful the new organization
will be.

While some factory managers I spoke with were initially skeptical of
what they characterized as outside meddling, most that I spoke to seemed
to ultimately appreciate the work of the Accord, and the union leadership spoke of it in even stronger positive terms. The reasons for its success
are manifold. First and foremost, it possesses the high-level enforcement
power I've been harping on throughout this chapter that is necessary for
change to occur. When brands sign on, they are *legally bound* to with-
draw from factories that do not pass the Accord's rigorous inspections.
If they do not, the unions—the other parties in the agreement—can
bring them to arbitration. Both parties, then, feel real pressure to change.

If the brands do not support factory reform (by paying for the improvements), they have to find another supplier; and if the factories don't make the changes, they lose vital business. Brands also don't run the show, as they do with their own codes of conduct. With 50 percent of the board of Accord being nonindustry members (i.e., labor unions), the workers are actually represented when standards and other policies are created. The equal governance means policy is not overly swayed by industry.

Rob, even in his role at the top of the pyramid, is also an executive of a different color; rather than a vested monied interest, like other CEOs who might own factories or be involved in the government, he is appointed by the ILO. That said, many cite Rob's forceful (I would say "refreshingly blunt") personality as a key factor in the Accord's might. As I talked with him, it was clear Rob wouldn't put up with anyone's BS excuses for why something didn't get done. Going after a factory to be sure it installs the right fire doors with regular check-ins might be unprecedented in this context, but isn't out of scope or reason—there's an enforced agreement to back it up.

As the Accord exemplifies, the relationship between policy and brands is key to the success of any major industry change. To uphold their lofty claims, brands should not rely on self-regulation, but instead should end the disincentives for countries to develop strong, enforced labor and environmental protections. Brands have a huge role to play in these countries and could be a powerful voice in saying, "We need you to develop your own labor and environmental regulatory infrastructure and we will not run away when that means that our costs will go up." And they can stop lobbying for trade agreements that do not include labor and environmental protections.

That's the aspiration, but so far in the conversations I have had, brands are applying some rather backward logic for why they should not get involved in government policy or even something like Accord. The enforcement bit, which Rob has helped bake into the legacy of Accord, is the problem. They've been making, changing, and ignoring the rules of the game without anyone saying boo for years, so to suddenly have an unsolicited referee on the field is unnerving. According to one major brand representative I spoke with:

> Brands can help government if government asks for it. Brands are responsible for their factories only. It's government's responsibility to look after factories. If you have power, you can't impose, you can only share. Threatening doesn't work well. If you threaten people, they [the government] might do that but they might not accept it in their heart.

What she's illustrating is the twisted sense of power dynamics that has gotten us to where we are today. In classic hot-potato style, the brands say, "The government is the only entity with the power to create better rules," and yet this ignores that the government is hugely disincentivized in this global regime *by the brands* from creating those rules to begin with.

So how do you presume a brand would see a government stipulation for certain building requirements (or wage requirements, or time-off requirements, or protections for childcare)? A warm fuzzy hug or a threat? Brands with deep pockets often have more power than local governments.

Giving power back to developing governments is a centerpiece of Rodrik's vision for ending the era of neoliberal hyperglobalization. By enacting stronger trade agreements, with language that provides for social and environmental regulations, we'll have legally binding policies that prevent importing countries from exploiting exporting countries. And to make them legally binding, Rodrik advocates for an expansion of government to pay for the enforcement of these policies: "If you want markets to expand, you need governments to do the same," he says. As someone into fashion, you might not think that you have anything to say about international trade. But if you don't tell your legislatures that it matters, this status quo of failed industry self-regulation will prevail.

While a government policy (through trade agreements or domestic laws that extend their reach across borders for access to domestic markets) would be the best possible external change, we could also achieve some gains by demanding transparency from brands in what they're currently doing. I don't mean "transparency" as CSR folks mean it on Instagram. Brands can not only name their factories, but also disclose the results of their audits and actually make sourcing decisions based on the outcome of those audits. Taking a page from the success of the Accord, brands should be responsible for paying for audits, too, and have a policy (one that they actually follow) that if factories are found to not be in compliance, they terminate those relationships if they do not come into compliance after a certain reasonable time. And, while we're at it, transparency on the wages of the workers producing for the brands would be helpful, too.

In the process of actually looking at numbers, it might become clear that something else is missing as well: transparency itself within a brand's own operations, thanks to everything from middlemen like Li & Fung and auditing firms to nonalignment between CSR and sourcing. When companies use these third-party facilitators, reality gets further and further away from them. By maintaining a more direct relationship with suppliers, just like our cotton farmer Carl Pepper imagined, brands can and should develop a more coequal relationship with an open discussion about how the prices they demand affect the way the factory operates and how much workers get paid.

None of this is particularly hard or devastatingly expensive work, unless you factor in the potential loss of business when customers see the truth of a brand's practices. As a Sri Lankan factory owner I spoke to asked, "Are we trying to solve sustainability or are we solving a PR issue?" Shifting the terms of the dialogue is entirely in our hands.

＝

The economist Joan Robinson said, "The misery of being exploited by capitalists is nothing compared to the misery of not being exploited at all." Forty years ago, when the promises of globalization shone bright above our magnificent "city on the hill," the apparel industry seemed to offer a way up and out of poverty for countries and people the world over. And indeed, as we've seen in China, Bangladesh, Sri Lanka, and elsewhere, it has been a part of raising national incomes and some metrics of quality of life.

But as we have also seen, those headline figures are far from the whole story. The current system is destroying the planet, ignoring the losers, and creating precarious jobs with precarious futures. Many hold on to this vision even now as sweeping changes to "the system" are proposed within the industry and society at large—a dismantling of neoliberalism and the extreme capitalism that it has created, which has deregulated social welfare programs and made individuals see every part of themselves as a way to earn capital. That vision of the future involves a return to a truly democratic capitalism, where the market meets the needs of the people rather than creates unnatural competition among businesses because of some dusty doctrine that the market knows best. If we relinquish some of the marketplace's power over our lives, the extreme neoliberal capitalists say, then millions of people will be thrown off the "escalator out of poverty." In their eyes, it's better to earn $95 a month, or call twenty-four-hour shifts "freedom," or have your backyard be full of trash and fabric scraps, than to have no job at all, or to be a starving farmer whose crops don't have a chance of growing on a floodplain.

But is this really the highest bar we can set for ourselves? Putting in place policies and laws that protect the environment and workers' rights beyond the bare bones of existence isn't unprecedented. Labor activists of the early 1900s fought for, and won, protections that we must extend to laborers around the world.

"We are not subject to change, we inform change," said an ILO representative I spoke with in Bangladesh. "Transformations [to quality] take time. If there are changes, [like] shifts to consumer patterns in Europe, maybe there [are] different types of buying. Maybe jobs will be lost but other jobs will be created. You can't say we will continue to destroy the world to keep this model." My understanding of these poignant words is a little less poetic, but hopefully as effective: We need to step up and use the power of our knowledge to create clean industries and actually lift up others, even if it means brand CEOs make a bit less, or we pay 25 cents more for our jeans. As Ashila in Sri Lanka told me, she did so because "when something is unfair, there is an anger in [my] heart that comes from within."

Five

Reclaiming Essentials for All: Packing and Distribution

Laura started her career working on the corporate side of a major airline. She enjoyed her job, but feeling her calling was elsewhere, she enrolled in an associate's degree program for wildlife biology. Three quarters of the way through, she got pregnant and never finished her studies. Three kids, one divorce, and twenty years later, she found herself looking for a new job with no résumé to speak of and none of the skills that are required for high-earning jobs today.

Then she stumbled upon what she thought was an employment pot of gold. Amazon, near where she lived outside Seattle, was hiring. The skill and knowledge requirements were minimal, and it had great health insurance—a necessity for a single mom. On top of that, when she got a reply to her application, she only had to come in for a "hiring visit," and the only real "interview" she had was a drug test. She walked out of that first meeting with a company badge and got called with a start date within a few days—the prize for passing the drug test.

Her new job was nothing like the work she'd done in the past. It was physically very challenging and mind-numbing, and once things got really busy at the end of the year—a two-month "peak" period when she had to be at work at 6:30 a.m.—it was impossible to drop her youngest child off at daycare *and* get to work on time. If she used her quarterly allotment of unpaid time off (UTO) to make up for the half-hour gap between the earliest day care drop-off and her clock-in time, she'd eventually run out of UTO and would likely be fired.

Over time, as her aches and pains (literal and metaphorical) worsened, Laura learned she wasn't the only employee with this problem. Many of her coworkers were also parents, yet there were no childcare accommodations for them, during peak or any time of year. According to federal guidelines, low-income families (a group Laura and many of her colleagues fell into) should not spend more than 7 percent of their income on childcare. For Laura, that percentage was more like 70 percent. Indeed, since 1960 the USDA found the cost of childcare has risen 800 percent nationwide, along with increases in lots of other essentials like housing and healthcare. But when Laura made the case for a childcare subsidy with her site manager, which would alleviate this huge burden of so many of her coworkers, the answer she got was "I don't know about that," and nothing came of it. After I reached out for comment, Amazon pointed to a program that launched in June 2020 that, according to Amazon's website, will run until January 2021, which provides a maximum of ten days of subsidized, emergency backup childcare for $25 a day in an in-center childcare facility or $5 an hour for in-home care. This narrow emergency measure does not appear to address the issue that childcare just was not open in time for Laura to make her shift.

Amazon is America's second-largest private employer, after Walmart. We've landed here in the journey of our jeans because today it is very likely their next step after being cut and sewn. While Amazon might be known for its essentials like toilet paper, it's actually the largest apparel retailer in the country, measured by the number of shoppers. It only just snatched the title from its longtime competitor Walmart. As consumption has shifted from physical stores to digital ones, Amazon's might in the clothing industry has only grown. In 2019, online commerce beat out physical stores in sales volume for the first time. And more than one third of online clothing commerce happens on Amazon.

Since the COVID-19 pandemic, the divergence between physical and digital has only escalated: US online sales in June 2020 were up to $73.2 billion, 76.2 percent higher than the year before, and experts predict that these behavioral changes will stick even in a postpandemic world. As former fashion stalwarts like J.Crew, Brooks Brothers, Ann Taylor, JCPenney, and DVF—brands with a physical and online presence—have sought bankruptcy protection from the effects of COVID-19 on the retail industry, the future of the fashion retail economy is looking a lot more like an Amazon economy.

Amazon makes money from fashion in various ways: (1) by serving as a traditional retailer, where it buys garments at wholesale prices from a brand and sells them retail on its site; (2) by selling clothes under its own private label (in these two instances, the manufacturer sends the products directly to the Amazon fulfillment centers); (3) by having third-party companies sell on its platform—in this instance either the third-party company or Amazon does the fulfillment; and (4) fashion companies can sell products on their own sites (not on Amazon.com) and then pay Amazon to fulfill the orders. Amazon is becoming the infrastructure of commerce—selling things, competing against others selling things, all while owning the platform where most of the deals are done. It's partly for this reason that Amazon has become a leading target for congressional hearings focused on unfair competition.

While Amazon's clothing bread and butter are in basics (tees, tanks, socks), it has been working to dig its nails into high fashion for years. In 2011, it launched a flash-sale site à la Gilt that hung around until 2016. In 2012, Jeff Bezos, founder and former CEO of Amazon, cohosted fashion's biggest night out, the Met Gala. Meanwhile, Amazon cosponsored the first New York Men's Fashion Week with the Council of Fashion Designers of America (CFDA) from 2015 to 2016. In this period it also launched Amazon Fashion, a dedicated online fashion platform, and expanded its own private label offerings, which now comprise more than

22,000 individual items—across all products, Amazon has more than 111 labels. Half of Amazon's private labels are in the clothing, accessories, and footwear categories. In spring 2020, editor in chief of *Vogue* Anna Wintour announced the media property's partnership with Amazon Fashion called Common Thread: *Vogue* x Amazon Fashion, a digital storefront geared to supporting American designers and brands hurt by the pandemic. Bezos had arrived to save the fashion world—by taking more and more ownership of it.

Since Amazon is so entrenched in the sales and distribution, and even now design, of our clothes, it's a key place for us to focus our attention as our jeans make their way into our closets. Along the way, we'll see how Amazon's way of doing business has significant implications for the fashion industry, as well as for all of retail and the global economy at large. "When we talk about Amazon, we're really talking about the future of work," Stuart Appelbaum, president of the Retail, Wholesale, and Department Store Union (RWDSU), told Recode's Shirin Ghaffary and Jason Del Rey. "Other employers feel that if they want to survive, they have to find a way to change their working conditions to replicate Amazon." RWDSU represents thousands of retail workers employed at companies like Macy's, Bloomingdale's, and H&M, along with workers at household American brands like General Mills and Coca-Cola.

So far on our journey, we've looked into the conditions that make it hard for the millions of people who make our jeans to live healthy, fulfilling lives around the world. Now, as we'll see, we must add to those people the workers responsible for getting them to our door. At Amazon, that's more than 250,000 such workers in the United States, who spend their days at the fulfillment centers (an ironic term if there ever was one). Their median annual salary is $28,446, a wage that forces many Amazonians to go on government assistance, including food stamps— which, Laura pointed out to me, were not even accepted at the Amazon cafeteria in the center where she worked. Our tax dollars support people who can't get their basic needs met by working at one of the world's most successful companies, yet Amazon itself managed to not pay a single cent in federal taxes for 2016–19 filings. And sitting at the top of it all is Bezos himself, who became the first person on the planet to amass a fortune over $200 billion, hitting $204.6 billion in August 2020, and substantially above Bill Gates, Bernard Arnault, head of the LVMH Family (the owners of Louis Vuitton, Céline, and Fenty, among others), Warren Buffett, and Mark Zuckerberg.

The coronavirus has only escalated the wage and welfare gap between Bezos and Amazon's warehouse employees. You may have heard (or even participated in) some of the insane online shopping frenzies that began once COVID-19 started its path of destruction through the United States in early March 2020. Toilet paper, among other "essentials," was out of stock everywhere; with physical stores closed and "stay at home" orders in place around the world, buying things online became a principal way people could get what they needed (and wanted). Amazon's stock prices skyrocketed as a result, increasing by about 50 percent in the first nine months of 2020. Bezos's personal gains also skyrocketed: By mid-2020, his net worth broke the $200 billion mark.

The fact that Amazon has control over so much of what we need (and want) and the convenient, touch-free ways we get it meant that Laura's former job processing inventory at Amazon, as well as those of all the other folks in a similar position, was suddenly elevated to "essential," too. Tear-jerking Amazon commercials played on TV, reminding viewers that "delivering the things people need has never been more important," and thanking the "Amazon retail heroes" who made that possible. Other ad spots showed off smiling employees whose children were proud of their parents' brave work in the distribution centers. Dark navy Amazon Prime delivery trucks haunted the streets of the Minneapolis suburb where I spent the first few months of the pandemic, its smiling logo a reminder that my and my neighbors' essentials (including sweatpants, which had quickly replaced the jeans we had heretofore been living in) were on the way.

Meanwhile, as we will see, something quite different was happening inside the fulfillment centers, where workers continued to toil for the salary that barely covered their own essentials, while putting themselves at risk for illness by going to work at all. Amazon did make some gestures toward improvement during the outbreak—a base pay raise of an extra $2 an hour, double overtime pay, and unpaid leave were granted with no questions asked—but those benefits for the "essential" work disappeared as some states appeared to get the virus under control. Ghaffary and Del Rey reported in June 2020, seemingly in response to public criticism that Amazon paid a bonus of $500 per full-time Amazon employee and $250 for others.

The treatment of Amazon workers has become such a big deal because of the retail-is-king trajectory our economy has been on. As we discussed in chapter 1, not too long ago, the United States shifted away from its roots as a manufacturing economy toward a retail economy. Manufacturing had held steady in the United States from the mid-1960s until 2000, when over the course of the following decade the number of manufacturing jobs plummeted by one third. During the same decade, retail and online shopping sales increased by one third, excluding a temporary dip after the 2008 recession.

Part of this transition to a retail economy involved the rise of merchant brands we've discussed—Levi's, J.Crew, and Gap—which moved production out of the United States and lowered clothing prices, so we went shopping more, leading to an increase in retail jobs. But the more recent stage in the retail evolution, from physical to digital, is worth scrutinizing. To meet this rising digital demand, online companies, principally Amazon, have created hundreds of thousands of warehouse jobs since 2010, adding workers at a rate that was four times that of overall job growth, yet the Economic Policy Institute cautioned that despite these headline numbers, Amazon actually created *no net gain* on the overall number of jobs in the areas where they do business. Despite these findings, Amazon has collected at least $1 billion in incentives from state and local governments to open fulfillment centers under the premise of job creation. The graph below compares employment at Amazon retail versus brick-and-mortar stores. (We'll get to the tail end of the graph, where the "Amazon economy" shows a decline in job growth around 2019, in a bit.)

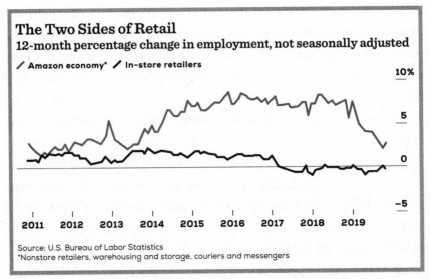

The Two Sides of Retail
12-month percentage change in employment, not seasonally adjusted

Amazon economy*   In-store retailers

10%
5
0
−5

2011 2012 2013 2014 2015 2016 2017 2018 2019

Source: U.S. Bureau of Labor Statistics
*Nonstore retailers, warehousing and storage, couriers and messengers

Job creation is great; however, as a brand and as an economic super-power, Amazon is shaping the future of our world in the way it likes, and that sits at odds with a healthy society. If it continues on its path of world domination, we'll see fewer jobs and fewer protections for people who hold those jobs. But if we use our power to speak up now, we can have a very different future. Whether we choose to deem that "essential" enough to change the way we vote and shop is up to us, and the best way to do that is to see for ourselves what essential really looks like today.

WHAT ESSENTIAL WORK REALLY LOOKS LIKE: INSIDE AMAZON

"Work hard. Have fun. Make history." These were the words that greeted me as I stepped into the Amazon fulfillment center in Edison, New Jersey, one of 110 such centers in the United States. I was there for a public tour of the facility. While it was geared toward impressing the public, starting with my overly bubbly tour guide wearing TGI Fridays–style pins on her uniform, chirping uplifting fun facts the whole time, I was privy to some of the experiences workers encounter every day according to the three employees (two former and one current) I spoke to. Including that internal motto, which I quickly snapped a photo of before tucking away my phone. As Laura and my tour guide explained, there's a "no phones allowed" policy on the warehouse floor; employees stow their phones and other valuables in lockers prior to going through security when they arrive each day. That means if they want to check their phones (social media, email, or potentially urgent messages from kids and relatives) during the day, they need to exit the security checkpoint, then get recleared, all in the time allotted for their breaks—somewhere between fifteen and thirty minutes.

My tour at the Edison facility began after a typical morning shift starts, which on nonpeak days is seven a.m. I'd enjoyed a relatively calm commute from Brooklyn, but the associates I talked to explained that many of their coworkers live upward of one hour away from their respective fulfillment centers. "There's no being late," Laura emphasized, so folks with a longer commute might require a 4:30 or 5:00 a.m. wake-up, accounting for traffic in the large metro areas and the lack of affordable housing near the warehouses. Becky, a twenty-five-year-old former associate from the Midwest, was grateful that she lived relatively close to her facility; she didn't need as much of a buffer to get to work and find a parking spot near the entrance without being late. Considering the size of some of the facilities—the Edison center is a 923,000-square-foot former Frigidaire plant (a literal manufacturing-retail flip), the size of sixteen football fields, my tour guide cheerly told us, with 2,500 full-time employees—where you park actually makes a difference. Walking across a giant parking lot, stowing your things, getting through security, and then walking through a giant facility to finally arrive at your station and start your workday takes significant time.

If for some reason the daycare, traffic, parking, or leg gods aren't smiling on you one day, and you scan your badge to check in after seven, you might get "dinged," one of the many unofficial warnings employees can receive for any sort of infraction. Employees can dip into their UTO, if they are late, but it provides only a thin cushion. If you take more than twenty hours of UTO per quarter, which adds up quickly if you're fifteen or thirty minutes late daily (as Laura would have been to allow her to drop her kids off at daycare), you would be called to HR and may be terminated.

According to Sam, an associate who currently works at the same facility Laura did, the actual workday starts off on a quasi-motivational foot. Teammates gather for a "stand up" meeting with the area manager, who leads them through a few stretches to warm up. Working at Amazon is not yet considered a team sport, but it might as well be for the intensity of the physical labor it demands. Depending on your facility and role, you could be lifting ten- to forty-pound items (like large bags of dog food), squatting, climbing, reaching, standing in one place, and/ or walking, walking, walking, across all those football fields of space. Between warm-ups, the manager chirps out motivational tips: how to be more efficient, how to use a "C-grip" when handling items to prevent hand and wrist injuries, and above all, hammering home the necessity of high-quality work.

The divide between Amazon execs who approve these scripts and the people who execute their vision for efficiency is clear in this small but telling example. The C-grip, as Sam explained, is "technically possible and true and would help" with the cases of carpal tunnel and other pain that Laura, for instance, had to get two surgeries (one on each arm) to remedy. But given the unpredictability and variety of sizes, shapes, and weights of the boxes and items employees need to handle, the C-grip is "out of step from the real world of how the work happens. Because you can't really adhere to any of that stuff if you want to go fast"—and you have to go fast if you want to keep your job. Workers are all aware of this but rarely speak up; there's no point in arguing with management, since they have to encourage the "safe" method even when they themselves know it's incompatible with productivity requirements.

There are five or six main roles at a fulfillment center that fall into two buckets: inbound and outbound. Inbound workers are called "stowers," and they're responsible for processing items—like our jeans coming from Bangladesh—that arrive at the center into Amazon's system. Amazon fulfillment centers are mostly named after nearby airport codes—Edison's is LGA9, after LaGuardia airport—a nod to the fact that things are coming to them via plane and train. A staple product like a basic jean would get shipped, while a fashion denim would likely arrive by air to receive goods before trends pass.

Stowers receive items at their station, which is equipped with a screen, a stepladder, and a yellow scanner. The scanner is an Amazon associate's magic wand—or handcuffs, depending on how you look at it. It keeps track of every move they make, literally, in order to collect data on how efficiently this essential work is being done, individually and collectively. In modern automated facilities, there are stower stations every six feet or so positioned around a massive fenced-off area filled with a field of mobile "pods," where the stowers stow the freshly delivered items. Call it an early form of social distancing, or solitary confinement, since the distance between stations prevents colleagues from being able to communicate with one another while working.

Pods are innocent-looking enough—portable square aluminum frames on tall legs, wrapped with yellow Kevlar cloth in order to make cubbies, pockets, and shelves on all four sides. Like a loyal pet, they practically greet stowers at the door as they arrive at their station in the morning. Laura's pods arrived by way of a large orange Roomba-like robot that is officially called a Drive Unit but associates refer to as drivers. (Of the 175 Amazon fulfillment centers worldwide, currently, 26 use these robots.) The driver wedges itself under a pod, picks it up, and delivers it to a workstation for the workers to start processing. Watching the pods being driven around the enormous enclosed space by a small robot in Edison was like something out of a futuristic movie—a completely automated internal delivery process. Unlike physical stores' inventory systems—shelves or racks with like items arranged together by size and color—pods are organized randomly by design. It's much easier, for instance, to quickly spot a stick of deodorant next to a pair of jeans, rather than trying to find a size 29 indigo boot cut jean in a whole bin of identical-looking denim. But the system also reflects the excessive consumerism that Amazon enables. Sam recalled pods of his recent past: "A set of extra large jeans in a plastic bag are crumpled on top of a sex toy, which crushes a bag of potato chips, filled at the sides by Apple gift cards, Blue Lives Matter patches, Mrs. Meyer's soap, and a stack of *White Fragility* in paperback, all on top of a heavy pack of dog food." It's a *Mad Libs* of trends. Certainly, it was predictive of the zeitgeist of the months after we spoke, in early 2020.

In the automated facility I visited, the stower does not have to figure out where to place a product in the disarray of the pod. Instead, the stower scans the code on the product and a field on the pod lights up, directing them to the optimal place for stowing—inbound paint-by-numbers, on speed. The scanner literally controls every move a stower makes—retrieve this case of dog food and place it in that bin in that spot—a task that requires neither cognitive input nor even linguistic fluency, since there's always a photo of the product on the scanner's display. It also automates the work of the efficiency engineers I saw in Dhaka by tracking the stowers' scanning rate. If a worker is not actively scanning for two minutes, the screen starts counting "time off task," which can accumulate toward UTO and a write-up. Laura pointed out, though, that she and her colleagues were often reprimanded for time off task even during the rare lulls in activity when there was nothing to be scanned.

But I digress—let's get back on pace! Once you or I click BUY on our jeans, the next part of the process commences: picking, the first part of outbound. In a facility with robots, the driver will deliver the pod full of randomly combined things to a picker's station to retrieve our pants, climbing up and down a little sliding ladder to reach them if they've been stowed up high, or squatting low if they're buried down at the bottom of the pod. They'll scan the jeans, put them in a tray, and send it down a conveyor belt. Over and over and over. Pre-robot, or in a facility without robots like Becky's, the picker would actually have to walk around the entire facility to the various pods or racks, lined up like aisles in a massive store, retrieve the items, and bring them to their station to be scanned and binned. Doing this, Becky said, she walked an average ten miles a day. Her record was fifteen.

To finish things off, all of the items in our order—our jeans, favorite coffee, maybe a "self-care" treat we saw our favorite influencer use on Instagram—make their way, via conveyor belt, from the pickers' bins to the packers. Packers stand at the ready with their tools: a screen that operates as the brain of the operation—flat boxes of various sizes, a roll of tape, and a roll of plastic that gets filled with bursts of air for stuffing. As the products arrive, the screen shows the packer which size box to use and dispenses premeasured tape. In the space of a few seconds a package is made, with virtually zero mental engagement by the packer.

Then it continues down the conveyor belt, where another machine called the slammer slaps a shipping label on the package at a rate of eight orders per second. The "system" determines where the box goes next— to a sorting center or a delivery station. When it's ready to get on a truck, another worker will "stack walls of boxes like a bricklayer, so the truck layer is packed up with smiley face boxes," said Sam. "Unloading trucks is like Jenga," since the boxes can move around a lot in transit and get loose. And just like in the game, there's a potential for everything to fall down—Sam knows of people who've gotten concussions from falling boxes.

After a whirlwind trip through the warehouse, the packed boxes are ready for their next destination.

FROM HANDPICKED TO DRIVER:
THE AUTOMATED FUTURE OF
ESSENTIAL WORK

In the span of the twenty-four hours in which I took a break from working on this section of the book in July 2020, Jeff Bezos's net worth increased by $13 billion. That is what he alone netted in a single day, not Amazon total. An Amazonian working full time at the lauded $15 an hour would have to work for more than four million years to make the same as their boss.

Now, I'm not suggesting that this pace of growth is going to hold forever, but it's more of a reminder that we're dealing with serious inequality at the highest level of the fashion industry. What's making the problem worse is that the gap only seems to be getting bigger thanks to Amazon's unofficial but quite obvious response to the problem of mistreating its human workers. On the surface, there may be outward displays of care, or at the least appeasement: the C-grip speech that fails to protect workers' health, the ironic insistence on calling warehouse workers "associates," the motto that greets workers at the entrance, or the publicized temporary "pay raises." But underneath, what's driving Amazon forward—literally—is automation. Machines have been menacing Laura, Becky, and Sam; their "essential" work is increasingly less essential to Amazon's profitability.

I saw this happening before my eyes on my tour of Amazon. For example, the robot drivers. According to Amazon, drivers are ostensibly here to make things easier for human employees by eliminating a lot of one type of physical labor from the job (walking). Remember that Becky would have to trek across the whole warehouse all day to do her job as a picker, whereas drivers can deliver full pods right to an associate in an automated facility. Instead of walking, though, fulfillment workers now crouch down and climb up a ladder for close to ten hours a day. The work isn't easier, just different.

Amazon also claims that by adding these 200,000 robots, more human jobs will be created, and points to the 300,000 jobs that were added during the robot rollout. Sounds awesome, right? Until you step back and realize that those extra jobs are only needed to meet faster fulfillment and our increased demand. The introduction of the robots and other automations are *not* the job creators Amazon claims, but are actually behind the decline in the "Amazon economy" in the graph on page 124—because even as demand grows, if more machines get things done faster, Amazon needs fewer people.

The drive toward automation doesn't end with the drivers. Until recently, work like what the pickers do that requires evaluating the contents of a random assortment has been solely the domain of humans. Machines weren't able to do that kind of recognition, despite the mechanical nature of the work itself. But a new German company has recently unleashed a robot onto the market, Covariant, that's capable of executing such tasks at the speed and efficiency of people. It's only a matter of time before Covariant and its inevitable cousins make their way into the Amazon distribution centers.

It is not just that machines are replacing people at Amazon. The company is also pushing people to become machines, even blurring the lines between how the two are talked about. "Just like workers, the machines need to be recharged," my tour guide said, pausing before a driver that had pulled over on the robot highway. But from what I heard from Sam, Becky, and Laura, human workers are not actually being given that luxury.

Remember how the instructions from the scanner and automated pods made the work of stowing pretty, well, robotic? Sam told me bluntly that they didn't help. They "made the job worse, because it's both faster and more ergonomic than ever, allowing for and requiring constant motion without small pauses, and no thought whatsoever. See, grab, scan, put, button, see grab scan put button seegrabscanputbutton-seegrabscanputbutton . . ." his voice blurring in a way that made my own heart start to race on the other end of the line. And that is the job—standing, scanning, stowing, scanning, "eighty to one hundred items an hour," estimated Becky, with no opportunity to sit on the side-lines and recharge.

The mechanization of industrial work in the fashion industry has been going on for centuries, and what we're seeing now is just how far and wide the effects are on people when that robot expectation is taken to scale. When items need to be processed in less than one second, it's easy to see how a machine might actually be preferable to workers' mere mortal bodies. Humans may struggle or complain about their scanners' incessant demands for more, cannot carry a whole pallet of goods at once, and have muscles that get sore and bones that break and lungs that can get sick from coronavirus. Robots, on the other hand, don't get cranky over low pay (they work for free!), never show up late (they sleep at the office!), and cannot cry when they're yelled at. They are far less of a liability because they have no real needs.

Needs like bathroom breaks, for example. At the time of our conver-sation, since he's cross-trained in different jobs, Sam worked in another support role, where he essentially relieves people for bathroom breaks. I was curious about how anyone could go to the bathroom when their every second is monitored, with boxes and packages and robots and screens literally coming at you from all directions, similar to the con-cerns I had at the facilities in Dhaka. In stow at her facility, Becky said, it was easier to accommodate one's natural urges, since you could situate yourself with your tote close to a bathroom, scan an item, run to relieve yourself, then scan again in under two minutes (the cutoff for UTO). However, if you're in pick, and in a facility without robots, you could be sent all over the building at random, and it could take ten or fifteen minutes just to get back to your cart after retrieving an item. Amazon disputes this and says its workers are "never restricted from using the restroom."

The perhaps obvious alternative for mid-shift bathroom breaks—going on your actual break—is just as much of a risk, if not more. The Amazon fulfillment employees that I spoke to got an hour break total all day, two fifteen-minute breaks (one mid-morning and one mid-afternoon) and one half-hour break for lunch on their ten-and-a-half-hour shifts. But *run* is not just an expression in this case if you actually want to run to the bathroom in that time. It could take five minutes just to walk to and from the bathroom. (Amazon says restrooms "are a short walk away from each work station.") The breakroom could be just as far (Laura's was a mile and a half from the farthest corner of her warehouse), which leaves just five minutes after the two-way commute to actually "break" between scans. And if you want to check your phone during that time . . . ? Well, remember you'll have to schlep to the locker area back at the main entrance, go through security to leave, reenter through security, and haul it back to your station.

The same impossible time travel would need to happen to have a proper lunch break, meaning the consumption of food (another thing robots are exempt from). "God help you if you have to heat something up," Laura added, since there were never enough microwaves in her facility, and waiting for one could take up your whole break. Lunch is rarely a time for social nourishment, either. Laura was bitter about her managers' "harping on connections" among teammates, as they chirped, "It makes the day go faster when you have friends!" She said she would sometimes buddy up with someone from her hiring class, but not necessarily because of a true friendship. "I know you, you're familiar," she would think, waving to someone she recognized in a sea of downturned faces.

Even though it's one of the few times they're physically close enough to do so, coworkers rarely talk to one another during any of their breaks, especially not about the tiresomeness of their daily grind. Those "issues are out of our control," Laura said. Becky similarly dismissed the idea of even small talk happening among coworkers, let alone big talk about how to make their work lives better. To do so would be futile, since "it just reminds you of how helpless you are." Misery may love company, but when you only have 10 percent of your day to not be *in* the misery, you may not want to rehash it. "The depth of all this resignation is constantly shocking to me," said Sam. Plus, complaining too loudly would only make waves, and no one wants to be the one to poke the beast.

Despite all the difficulties, Sam told me that everyone, in general, at his facility wants to do a good job, because of or in spite of the gamification strategies Amazon has concocted. Scanners can also double as pseudo Nintendos when workers get the chance to turn their pace tracker into a race car, or another method of encouraging competition among employees. Getting the highest score may even enter you into a raffle (the prize might be a Snickers bar). These games seem masochistically silly, and patronizing, but some workers like them because it's some way to be engaged with what they're doing. Sam tries to look at the bright side of things in this way, and takes pride in his work; the alternative of dwelling on how there's no way out of the system "just makes it miserable."

Nevertheless, Sam told me about coming home so drained that the only thing he could think to do on his off days was watch TV. He was also physically exhausted, and sometimes needed to resort to taking days off without pay in order for his body to physically recover from a shift. Becky described how easy it was for her brain to wander while on task, since her higher thought processes "went to sleep" while she was scanning. The work requires "just enough [focus] that you really have to pay attention" but not enough to avoid "becoming a machine." When I asked her what she thought about while scanning and toting and walking up to fifteen miles a day, she replied, "They're never good thoughts." She could see the same thing happening on her coworkers' faces, whenever they, too, slipped into machine mode.

The expectation of machinelike output while providing so little control over one's workday isn't just an annoyance; it has serious real-life health consequences. Researchers have found that people in jobs without a degree of control over one's work are more likely to become depressed and experience severe emotional distress, even controlling for pay. And that just makes sense. As writer Johann Hari puts it, "You aren't a machine with broken parts. You are an animal whose needs are not being met."

I told Sam that I'd previously spoken to workers who make our clothes, like Rima in Bangladesh and Danu in Sri Lanka, who also confessed to feeling like machines. He humbly suggested his story probably had nothing on theirs. When I told him that in fact some of the challenges they faced sounded really very similar, there was an uncomfortable laugh from the other end of the phone, then silence.

Instability and Inequality

The personal and professional trajectory of an Amazon fulfillment employee is anything but certain. Most workers are in their twenties or thirties and physically fit, according to the associates I spoke to and what I saw on my tour. Anyone older seems doomed to fail purely as a consequence of biology. At just thirty-nine, after less than two years working at Amazon, Laura developed severe plantar fasciitis, carpal tunnel, and knee problems, conditions that she and everyone on her team knew could not be alleviated by the morning stretch routine that Amazon prescribed. At least Amazon offers a healthcare plan—"one of the really amazing things," Laura said—but the job is so "physically brutal" it actually requires making use of it frequently. Becky went through two tours at Amazon until her twenty-five-year-old ankles gave out past the point of no return. Imagine what the job does to older workers, like the woman in her late seventies whom Sam met at his site. "This is not how I imagined spending my retirement," she told him.

Becky is not the only one to have her job taken away because of the demands of the job itself. And it might not have happened if not for peak: the last two months of the year, when shopping, especially at Amazon, skyrockets for the holidays and lands in Laura's, Becky's, and Sam's hands. During the 2019 winter holidays, Sam's center shipped 26 million items across two months, and workers' hours increased from forty hours a week to fifty-five in mandatory overtime. Sam told me that the normal level of competition gets a boost during peak, as all the facilities vie for the top spot when it comes to orders shipped. Winning might be useful for the facilities' higher-ups, but not for workers like Sam, Becky, and Laura. Amazon raffles off things like an Echo or free lunch for top-performing individuals and facilities, but the reward is minimal compared with the downside. Once it's established how quickly workers can work during peak, that becomes the expected baseline pace for the other ten months of the year.

Peak is just one reason for Amazon's shockingly high attrition. Remember, the earlier start time required during peak eventually made Laura have to choose between her child and her job. Laura estimated the attrition rate was 80 percent a year at her center. During my tour, I noted a separate entrance for new recruits; Amazon's assumptions of high turnover are built into its architecture. After less than two years, Laura was the last member of her hiring class of twenty left. Sixteen had jumped ship after the first six months. Sam claimed his center had almost a 100 percent annual turnover of employees in pick. Their observations align with research that the average turnover rate for warehouse workers in counties with Amazon fulfillment centers was 100.9 percent in 2017, the latest year for which data is available. How do you even have attrition of over 100 percent? It means that the total number of people that left warehouse work is higher than the average number of warehouse workers for the year. If you do stick around, Laura added, she didn't feel there was a real chance of promotion after the three-year mark, since most people at the mid- to high levels—generally, those who don't perform physical labor—are too comfortable to leave. You can see how, with no light at the end of the tunnel, most people don't even last a year.

The pay also doesn't help. The workers I spoke with earned between $14 and $17 an hour. This was before Amazon raised its minimum wage to $15 at the end of 2018 in response to mounting pressure over working conditions. When Sam started, he made $14 an hour; when he got the raise to $15 per hour—a much-hyped media moment—he lost his stock options and productivity bonus.

That compensation doesn't go very far today for a single person, let alone if you're supporting other humans with biological needs, like Laura does. At the end of her tenure at Amazon, she made $17.25—a far cry from the $75 an hour she made at her former airline job, and even from the $35 an hour she calculated would make her comfortable, aka her living wage. With that income, she might be able to afford a modest three-bedroom apartment for her four-person family, buy groceries, pay for childcare, *and* even treat herself to a haircut or a coffee once in a while.

Part of the problem, which extends beyond Amazon, is that even when the economy was growing rapidly pre-COVID-19—if we're just talking about the traditional way of measuring prosperity through the GDP—costs for housing, healthcare, childcare, and higher education also increased significantly without matching wage growth. One reporter has called this crunch the Great Affordability Crisis. We can pause a moment to note that if the minimum wage from 1968 kept pace with productivity growth, it would be $24/hour, which puts the pay raise at Amazon the media went gaga over—to $15—into perspective. The average American, including those Amazon workers, earns less now than in 1979, after accounting for inflation. As a result, even pre-COVID-19, 17 percent of adults could not pay their current month's bills in full, and nearly half of US adults didn't have $400 in savings to cover an emergency expense. The pandemic and inadequate government response have created emergencies for many families, amplifying the existing instabilities of our labor force that have been forming below the surface for years.

By comparison, these low-income workers may be employed by some of the wealthiest individuals in the world (America's top earners live up to fifteen years longer than the lowest earners). Pre-pandemic, 30 percent more Americans were on food stamps than in 2008, yet we have twice as many billionaires as we did a decade ago. Mr. Bezos is ridiculously wealthy, but executives in general have been raking in more and more over the past four decades. In 1965, the average CEO earned $924,000 (about 20 times more than the typical employee), compared with $17 million in 2018 (or 278 times more than workers). The top 1 percent of earners are $21 trillion richer compared with 1989, whereas the bottom 50 percent are poorer.

The reason for this huge discrepancy in the worth of workers and their bosses goes back to the neoliberal concept introduced in chapter 2: shareholder primacy and the principal tool to achieve it, executive compensation. Shareholder primacy, as we discussed, says that companies should not focus on all stakeholders. Instead they should dedicate themselves to creating the most profit. In order to achieve this, corporate boards made more executive compensation tied to stocks, rather than paying straight salaries. The better the stock performed the more the executives made. Executives were then fully incentivized to move jobs to countries with cheaper labor, automate jobs that could not be outsourced (cue the orange Roombas), and depress wages of the remaining workers by impeding union formation. The costs were cut, the earnings increased, the stock price went up, and the pay gap widened. Amazon has in fact been on the leading edge of this model: It pays above-market compensation for its executives, but it does so by paying below-market salaries and making up the significant difference with stock.

While their boss sets his sights on colonizing outer space, his workers' very life expectancy may be decreasing from the combined stress of their jobs' inadequate pay and unreasonable conditions. Despite steady increases in GDP, life expectancy in the United States has decreased every year since 2014, with the exception of the most recent year's data; however, it's still a net drop from life expectancy in 2014. Some US counties now have even *lower average* life expectancies than Bangladesh.

What is driving declining life expectancy? A rise in what Anne Case and Angus Deaton coined as "deaths of despair"—a surge in deaths from alcohol, drugs, and suicide. "Something is making life worse," Case and Deaton write. What is that something? "Rising economic and political power of corporations, and the declining economic and politi-

cal power of workers," which leaves workers at the mercy of big companies on an uneven playing field. What would keep Amazon's and other
low-income workers healthiest is a reconfiguration of how they do their
work, including the political and economic power structures that shape
the entire labor market, so that our "essentials" don't cost other people
their own essential well-being.

GETTING STRATEGIC WITH IT: ENSURING THAT JOBS PROVIDE ESSENTIALS (AND MORE) THROUGH UNIONS AND INDUSTRIAL POLICY

The Americans who came before us worked hard to create the labor
advances that we celebrated just two chapters ago, so why are we dealing
with declining life expectancies in our own borders? As company executives worked to outsource jobs to countries without those labor protections, they also cut down those protections right here at home for those
jobs that could not be so readily outsourced, like Amazon fulfillment.
Amazon has made any talk of union formation verboten—except for
union bashing, that is. Becky put it in interesting, if not humorous,
terms (because if we can't laugh what's the point?), telling me that management "made you feel like an idiot" for bringing up the subject of
unions. As a way to dismiss the need for unions at all, the company will
show new employees a video about how working there is so great; Becky
shared that when the subject of unions came up with management, workers were shown testimonials from employees who lost twenty pounds at
Amazon from all the physical work, ostensibly arguing that being able
to exercise all day long is a perk. At Laura's facility, managers spoke about
how a union would "destroy my personal relationship with my employees"
through the insertion of this arbitrating middleman. As evidenced by the

helpful response to her ideas about childcare, however, that relationship isn't exactly being nurtured with love and respect. "Bitch, please" was what she had to say about that.

Indeed, managers didn't just try to convince employees that unions were bad. Workers feared severe repercussions for organizing. They wouldn't even speak of unions (if they had the energy at lunchtime) because they feared that everything could be seen and tracked. If she did speak out in any meaningful way, Becky worried that she could be fired from Amazon and blacklisted at its many subsidiaries. While Amazon spokespeople dispute the cause, there have been several reports of workers getting fired after speaking out about conditions within the fulfillment centers and further reports of Amazon monitoring the union-organizing efforts of their workers in Europe.

The viability of unions isn't a guarantee by any means, and not only because some companies will do everything in their power not to let any form. Laura described what is probably a typical assumption that bolsters companies' ability to inhibit unions: that unions weren't for people like her, but more for "machinists or railroad workers"—traditional "blue-collar" workers. Sam observed a similar attitude in his facility, which he chalked up to the individualist American ethos; his colleagues blamed themselves for where they ended up, rather than being given the space to see that the system is just not set up for their success. In today's economy, where the essential nature and dominance of those "blue-collar" manufacturing jobs are being transposed to other professions, and when we have proof that no one is immune to the forces of nature, the ideas around who needs unions and why need to change. "I think the state of affairs in our economy is that people are used to it being broken and not working for them," AFL-CIO secretary-treasurer Liz Shuler told Shirin Ghaffary and Jason Del Rey. "That's the way the economy has been for some time, and not just due to COVID-19."

We have very clear evidence that the ability to work with a collective voice yields results. Union members earn on average 13.6 percent more than nonunionized workers. At present, Amazon management is setting the rules for everything that happens inside its walls; allowing unions would force executives and shareholders to share some of the pie—in power and in money. Unionizing would likely offer real benefits to workers, so much so that one former unnamed Amazon executive broke ranks and told Ghaffary and Del Rey that "unionization is likely the single biggest threat to [Amazon's] business model."

An important facet of the antiunion ideas at Amazon has to do with the other pandemic that boiled over into the collective consciousness in 2020: racism. Exceeding the statistic of union versus nonunion wages above, unionized Black workers earn 16.4 percent more than nonunionized Black workers. Amazon has a significant percentage of Black employees—more than a quarter of its domestic workforce. (The overall US population is 13.4 percent Black.) Most Black employees work at the fulfillment centers earning far less than their corporate counterparts.

The overrepresentation of Black workers is not unique to Amazon fulfillment; it can be seen across the low-skilled warehouse industrial sector thanks to a long legacy of racism. Black people who were enslaved were the foundation of not only the fashion industry but also the global economy. Postslavery, Black people in America have faced discrimination and legal hurdle after hurdle. One effect of Jim Crow laws in the South was to limit occupations available to Black people to low-paid sectors, including farming and domestic servitude. New Deal legislation formally allowed for collective bargaining, a minimum wage, and a forty-hour workweek with overtime regulations, but deliberately excluded domestic, agricultural, and service work that Black workers disproportionately held because of Jim Crow laws. Then, stacked on top of those laws, redlining allowed housing discrimination based on race, which is a significant reason for the segregated cities we live in today. Although the practice is now illegal, its impact continues to be felt in lower educational funding in lower-income, disproportionately nonwhite neighborhoods, creating a vicious cycle of de facto segregation and underfunding.

Because of the institutionalized racism, Black people also took a harder hit from the early days of automation and offshoring than white people, in part because they disproportionately held the lower-skilled jobs in the first place. Since 1954, Black workers suffered unemployment at a rate twice as great as their white colleagues, and that trend has remained the same every year since; and those who worked remained overconcentrated in the lowest-paying occupations that offered limited hope for upward mobility. As Martin Luther King Jr. said at a union convention in 1961, "Our needs are identical with labor's needs: decent wages, fair working conditions, livable housing, old-age security, health and welfare measures, conditions in which families can grow, have education for their children and respect in the community."

This interweaving of racist laws, policies, and attitudes, which have been both explicitly racist and implicitly so throughout history, amounts to what author Isabel Wilkerson convincingly describes as a "caste" system. As ladders of opportunity have been kicked away, progress for America's Black community has been thwarted. One significant way to create ladders of opportunity is through unionization, which, as we saw, leads to meaningful wage gains and a narrowing of the race wage gap. By blocking unionization, then, Amazon not only hoards power for executives but also perpetuates structural racism. The other avenue for labor reform that we've also explored in this book is government-level intervention. In the past, Black workers indeed turned to the federal government to block discriminatory policies, in places of business and in unions, with mixed results. But Amazon has an answer to that, too, which negatively impacts all workers and disproportionately hurts Black people. Beyond conspiring inside its walls to block unionizing, Amazon, in concert with other companies, spends oodles of money on lobbying. (In 2019, Amazon spent more than $16 million on lobbying, and its budget has increased by almost 470 percent since 2012.) It's not just Amazon. *The Atlantic* reported in 2015, "For every dollar spent on lobbying by labor unions and public-interest groups together, large corporations and their associations now spend $34. Of the 100 organizations that spend the most on lobbying, 95 consistently represent business interests." Amazon also has a history of hiring government officials after they've left their agencies. What better way for a company to ensure that government officials protect their interests by blocking threats to their power than by dangling the promise of a future high-power, high-paying job? Rising political power and rising economic power safeguard against pro-labor legislation.

It's unfortunate that American corporations think it's a zero-sum game between labor and management. There is strong evidence that countries with more collective bargaining have greater productivity, higher rates of employment, and their workers have more resilience against contracting economies. For individual businesses, listening to workers doesn't just mean losing power, either. Knowing the obstacles your employees face can make them more productive, since, among other things, they're less likely to quit (which also means fewer resources are needed to train new employees). If, for example, a union had been able to communicate Laura's childcare dilemma to Amazon management, it may well have helped to find a solution for her and other working parents. She might have stayed longer, slowing the revolving door of Amazon attrition and, with that, increasing overall productivity.

With all the knowledge you now have about worker treatment around the globe, as well as the history of unions, I hope it's clear that the two go together like mom jeans and dad sneakers. And yet, it seems, many executives are willingly blind to how the system works and their role in it. In 2018, Google's former CEO Eric Schmidt tweeted about wanting to find a "unicorn for the middle class" that would increase wages (a "unicorn" in tech speak is a start-up company valued at over $1 billion). The internet response was a not-so-gentle reminder that the solution to the disappearance of much of the middle class will not be found in a fancy new start-up, but through the work that was done a century ago in significant part by garment workers. Rather than exerting more power over workers by exposing them to some new system or software, we must go back to basics. Jeans, tees, and unions.

———

We need to revive the labor movement, but it alone will not ensure that the people who get our jeans to our door today will still have a job tomorrow. Unions can help improve existing jobs, and stronger domestic laws and international trade agreements can ensure a more level global playing field. But they cannot ensure that there will actually be enough human jobs in the future. At the moment, we are discussing the unfair distribution of wealth, but we do need wealth to actually distribute, and we can't just take that part for granted. Remember chapter 2, where we went into a deep dive into China, and you were like, "Why am I getting a history lesson on China in a book about clothing?" Well, that history showed us why China dominates fashion production today: Its government invested in the sector.

Many successful industrial economies today, including China, Germany, and Japan, use this government investment—aka industrial policy—to develop nationwide plans that promote domestic industry through training and subsidies. The Chinese government today has its Made in China 2025 plan to pursue advanced manufacturing, using hundreds of billions of dollars in subsidies.

It's not unprecedented for the American government to play a similar role in industry and labor. New Deal policies put 20 million people back to work and lowered unemployment rates from 25 to under 10 percent in nine years. Today we need the same kind of visionary thinking and investment. Legislatures must examine the global economic environment, identify a plan to defend against risk, and join forces with universities, entrepreneurs, and the private sector to invest in exciting sectors that will help Americans thrive—like a high-speed electric train, green energy, or the medical advances that will cure cancer or COVID-19. We need to look past the tenets of the extreme capitalism found in neoliberalism that have gotten us to this state, and get back to the core values that can make America great. Enterprising individuals and innovative, safe, fair, and reliable work. An economy and society we can be proud of. Dani Rodrik and Charles Sabel have coined the phrase "good jobs" as an umbrella term for working conditions that provide people with these essentials—things like stability, bargaining rights, a living wage, and, yes, chances at upward mobility and career advancement.

The irony of all this is that many white working-class people in America—including those people who get our jeans to our door—vote against their own interests in this regard. Why? Some working-class white Americans feel understandable ire at not getting a piece of America's prosperity, but have falsely targeted blame at immigrants and racial minorities and vote for candidates who foment racism. White working-class people who were first impacted by globalization, as we saw in Texas in the last chapter, were less likely to vote for candidates who supported New Deal–esque social services, like retraining, that could help *them* get good, better-paying jobs. They were also more likely to vote for authoritarian values. In Europe, researchers found the same trend where areas that lacked support for democracy were more negatively impacted by globalization.

In reality, jobs held by white workers are not being taken by other workers in the United States. It's the higher-ups, and the very government officials they elected to save them from their station—mainly some rich white men who have their own status to enjoy and preserve at all costs. Rather than blaming people who may look different or speak another language, but have the same concerns and essential needs (not being met), the frustration at job losses, job conditions, and lack of income gains could be better aimed at the executives and shareholders who remain unwilling to share the wealth they have in significant part because of the workers—and the officials who do their dirty work at the government level.

"What we have to realize is that throughout history poor whites and [people who were enslaved] and then free blacks were pitted against each other, and that was used as a political tool," explains Nancy Isenberg, historian at Louisiana State University and author of *White Trash: The 400-Year Untold History of Class in America.* "I think one thing we have to realize about white supremacy is that it leads to an advantage to the elite to pit these two groups against each other," *Politico* reported.

Revamping our economy with good jobs as its foundation requires a collaboration between these fragmented parts of the workforce, but it also needs collaboration between public and private sectors—government, the public sector, and business working together to invest in education, training, and expanding job opportunities as part of industrial policy. If we are not able to come together at this higher level as well, even the most united, accepting, and cooperative workers will continue to be at great risk of being forgotten, downsized, offshored, or replaced by robots.

Why not have our government, which is elected by us, work for us and foster healthy competition with others for a better world? I offer more specifics in chapter 9 about how exactly to do that with the power of our voices as citizens.

———

Within the walls of Amazon, the have-nots motor around like robots, which will soon take their jobs and send them out into a bleak job market, frustrated and desperate for a savior in the form of familiarity. We've seen that many possibilities for our future have roots in the past—the securities of good jobs and unions—and so it's up to us to ensure that those memories are accurately and vibrantly jogged when we go to the polls to choose the representatives who will fight for the legislation we need.

This does not mean tossing out capitalism as a system—it has proven to be a very effective way to deliver a lot of good. As historian Sven Beckert argues, capitalism is and has always been an evolving system. If we want to preserve the benefits of capitalism, we must purge it of neoliberal extremism, which has led to fewer good jobs, preserved systemic racism, created an economy that works only for the few, and destroyed the environment.

Beckert offers an alternative for these conditions that alters our notion of what "growth" might look like: "The long history of capitalism testifies to an enormous reservoir of human creativity, and our collective ability to vastly increase the productivity of human labor. We can see this as a threat—we are running out of work, and people will be unable to find employment—or we can see this as an opportunity—future generations will need to work many fewer hours and we will still be able to maintain our wealth as a society." Amazon can still do its thing, Jeff Bezos can go live on Mars, and yet Sam and Laura and Becky don't have to spend all the prime of their lives inside a warehouse working like machines. In turn, people like them—like us—can start to use our mental, emotional, and psychological capital on things that will impact multiple sectors of our generation, and generations to come. As we'll see more in chapter 7, our minds need to be cleared of the clutter of stress about our livelihoods before we can think beyond our own basic needs. So by helping the workers of America—and the world—receive their basic needs, we will all be working hard, thriving, and making history together—the kind of history we want our grandchildren to study decades from now.

Six

More Is More:
Consumerism Goes Viral

> Just for you to know: I decided to suppress all my fast fashion apps. ☺

> So who convinced you to stop?

> I had time to think. I hate the world we are stuck in. And then realized H&M, Zara, etc. were over to me . . .

This is a snippet from a text exchange I had with my friend in Paris, whom I'll call Claire, one afternoon in early April 2020. I had to blink a few times to confirm what I was reading, as this was a most unusual message from her. You see, Claire and I were roommates during my gloriously free summer before law school in 2008, which I spent in Paris. There, I: (1) ostensibly learned French; (2) stopped blow-drying my hair with a round brush (that "done" look, I learned, was a key tip-off of an American); and (3) decided one of my new life goals was to have a closet like Claire's, where all the garments had space between them, enough that they could be seen individually, rather than smooshed and tangled in the overstuffed closet that I had back at home.

When I spoke to her for this book in early 2020, pre-COVID-19, Claire described her style as *elegante chic,* a phrase I could never pronounce with her melodious accent but could wholly corroborate. She's into classic styles, like cashmere and moto jackets, more eclectic accents like lots of glitter, as well as select trends, like the teddy bear coat she was wearing as we spoke. She confessed that she bought the coat on her phone, riding in a taxi, after seeing it on the Instagram feed of an influencer she followed, on the first day of the sale season. (France, unlike most of the rest of the world, has government-regulated sale seasons called "Les Soldes.") She also frequently found herself tumbling head-first down the rabbit hole of Instagram stories, checking out the latest posts and updates about new collections and sales and checking in with her "obsessions," favorite influencers and designers. Even if she was near a physical store, she would almost always choose an online purchase. And although she had refined her shopping tastes over the years, she was also a "victim of fast fashion," succumbing to the accessibility of cheap, daring, playful clothes after losing weight a few years ago and wanting to dress her new figure differently.

Claire is an embodiment of the proliferation of online shopping that, combined with social media, has given us instant gratification clothes. The list of places where shopping takes place these days is much larger than where Claire frequents, and certainly much larger than where I first got a sip of the shopping Kool-Aid in the malls of Minnesota: essentially, anywhere you can get a 3G connection. Shopping this way may be exhilarating, but even before going cold turkey on thoughtless fashion, Claire expressed a concern about her habits. "I'm not proud of my behavior," she told me, her voice shifting from animated to dejected in mere minutes. A few weeks later, that whisper of shopper's remorse, of the voice inside her that was aware there was something not right about this way of living, had exploded into a roar—along with many other voices that had been roiling under the surface of society far longer than they should have.

COVID-19 brought to light a swell of frustration, injustice, and messages of our collective desire—and need—for change on a number of fronts. One of many: our relationship to things. Suddenly, our homes were no longer just supersized storage pods but also offices, playgrounds, and yoga studios (if we were lucky), making the amount of stuff we had even more apparent. They had grown to accommodate our retail obsessions; the median-size living space per person in a standard home went from 507 square feet, the size of a generous NYC studio apartment, in the early 1970s, to 971 square feet per person in 2015. The time we spent online was not just for leisure, instantly gratifying shopping, and Netflixing: It was also time for school, work, doctor visits, and socializing. The jobs that financed our shopping may have been cut back or cut entirely, as 14.7 percent of the American population became unemployed at the peak of the pandemic job loss in April 2020—disproportionately Black and Latino workers whose unemployment rates still topped 17.4 percent for Black workers and 16.9 percent for Latino workers by the second quarter of 2020, compared with 10.8 percent for white workers. Our need to look *elegante chic* for dates, meetings, and even shopping trips was suddenly reduced to whatever would look presentable from the chest up inside a tiny square on a screen—and "presentable" was a loosely defined word, as I can attest to. Many of the things that once brought us moments of joy, happiness, satisfaction, and worthiness—things that many of us, like Claire, felt might not be making us really any happier—suddenly felt totally out of place, useless, and unnecessary in the "after times."

As the early reports from the effects of COVID-19 on the retail landscape show, Claire is joined by many others in reconsidering their shopping habits—and possibly for the long term. Data about the effects of COVID-19 on all aspects of life is limited and evolving, but one survey by McKinsey & Company of two thousand UK and German shoppers indicated that well over half expressed greater concerns about their purchases' and brands' environmental impact, and had made changes to their lifestyle that reflected greater sustainability. Durability of our clothing is also more important, as 65 percent of those surveyed planned to buy longer-lasting items, 71 percent are planning to keep what they have for longer, and 57 percent are open to repairing their clothes; there's also been renewed interest in secondhand garments, especially among younger shoppers. Thanks to Claire's and others' revelations, the disposable fashion machine that produced all those cheap, nondurable clothes we've seen being made in previous chapters may be forced to slow.

Indications of more conscious shopping, as preliminary and limited as this survey suggests, is no doubt a reason to celebrate, given all that we have learned about the system behind our clothes, but why this isn't our default set of consumption habits is a more noteworthy—and shocking—story. It's the story of marketing, which you may be a little familiar with if you've ever watched *Mad Men*. The ways we shop today—largely and increasingly online, and to a smaller extent in physical stores—are not a coincidence. Our consumption patterns are not the result of a natural evolution of fashion, the economy, or human behavior—organic consumer demand, as a marketer might say. Like most of what we've seen so far about how our clothes are made, how we, ourselves, shop for clothes is also part of a complex, at times horrifying, matrix of manipulation. We've seen the market forces that have contributed to the consumption of human and natural resources at a record pace and the rise of fast, even instant fashion. This chapter plunges a proverbial sewing needle right into the heart of it all. It asks the question so obvious we all overlook it: *why* we buy the new jeans in the first place. If the answer is to be happy beyond that instant serotonin high, then, it seems, we've missed the mark. Big time.

In the pages that follow, we'll take a deep dive into the psychology of consumerism, understanding what drives us to shop and how we can start to peel back the layers of manipulation that people, especially women, have been wrapping their bodies and minds in for decades. We'll not only explore how market forces, from the real-life Mad Men days (and they were some *mad*, i.e., crazy, men!) of advertising to social media, shape our sense of desire, but also talk with the people on the ground—average shoppers, stylists, and influencers—who are feeding into and being fed by the fashion machine. It's time to step inside a 360-degree mirror of shopping habits that will make you think twice before you click BUY. And hopefully, you'll come away understanding how to reclaim your desires, own your style, and find true happiness—whether it's an in-person hug from your best friend or a new (or new to you) pair of jeans.

BRAND NAME: THE ORIGINS OF CONSUMER MARKETING

These days, especially in the Wild West of the online marketplace, everyone's got a brand. You need one to stand out in the crowd of profiles who might be buying or selling the same stuff as you, and to fit in with the people you aspire to be and live like. The origins of the word "brand" reinforce how entrenched this idea of commerce—of individuals as well as services—is in our society. Among its earliest usages in Old Norse, around 950 AD, branding referred to a burning piece of wood used to mark an insignia of ownership on lumber and cattle. Its application became disgustingly cruel when people who were enslaved, in the Americas and elsewhere, were "branded," to show ownership as well as for punishment.

Fashion occupies a unique place within the history of branding. We can look to the era of Louis XIV—the "Sun King" who transformed France into its *très chic* self—for the seed from which the market as we know it today grew. When the Sun King ascended to the throne in 1643, Spain, not France, was the regional superpower. Spain defined what it meant to be modern. As Spain expanded its trade network around the globe, and acquired lands and gold through imperialism and colonialism, its pinched and formal and mostly black (the most expensive dye at the time, and thus a display of wealth) dress was considered the most elevated among European courts.

King Louis wanted France, not Spain, to be the economic superpower, so he developed his own economic stimulus plan and he used the fashion industry to get him there. Yes, the fashion industry has been a key to economic prominence from Louis XIV, to industrialized America, to modern China. He put the full strength of his court behind investing in the domestic textile industry, which would come to employ a third of Paris's workforce. The government ensured its competitive edge by organizing guilds—the early days of industrial policy—and banned any imported goods that could be made in France, what modern economists call protectionism.

Among his countrymen, Louis also took care to polish his own image as the representative of the new luxury brand France was becoming. A night at the opera was a chance for him to show off a daring ensemble, complete with his signature wig of curls and red pumps. He also circulated a series of fashion plates engraved with illustrations of French goods and garments, which could be passed around with a wink and a nod; the plates were accompanied by witty, sometimes naughty, captions. His version of our glossy magazines, TikTok videos, and Instagram posts.

But to use fashion to become the superpower would require something more. The engine of fashion would need to be accelerated. And to do this the fashion season was born. Jean-Baptiste Colbert, the king's finance minister, was instrumental in launching the two-season fashion calendar and the very notion that fashion would ever go out of style. This was not a stylistic choice, but more crucially an economic stimulus initiative to ensure that new styles were introduced at regular intervals, twice a year, year after year. The French fashion calendar was a sharp departure from Spanish culture, which took pride in its consistent look; by contrast, ever-changing French fashion meant that, by design, what you wore one season would be passé in the next. And now that Paris was the new fashion capital, France would be the economic beneficiary of all those outmoded clothes. It had a major impact. Colbert said, "Fashions were to France what the mines of Peru were to Spain." Who's calling fashion silly now?

This was the original "planned obsolescence," as the new fashion seasons demanded that people not only dress for beauty and the weather, but to meet now ever-changing social expectations. For a member of the French court, wearing summer '44 (that is, 1644) attire on a warm summer '45 day would have been unacceptable, rendering even the most luxurious clothing unusable. The manipulation of desires to promote economic growth had begun.

═

It's unlikely that you've heard of Edward Bernays—I hadn't before writing this book—but everything you own, including the sense of who you are, has undoubtedly been touched by this man's revolutionary application of modern psychology.

Bernays moved from his native Austria to the United States as a child, and was working as a press agent in 1914 when World War I erupted. Skilled at crafting persuasive messaging, he was brought into the US Committee on Public Information to craft calls to action for people to support the war effort. He was so successful at this work that he joined President Wilson at the Paris Peace Conference as the envoy for America's message of spreading the gospel of democracy throughout Europe, as well as to help shape Wilson's image as the defender of the free world.

"When I came back to the United States [after the war]," Bernays said in an interview, "I decided that if you could use propaganda for war, you could certainly use it for peace. And propaganda had a sort of bad connotation because of the Germans using it. So what I did was try to find some other words so we found the words 'Council on Public Relations.'"

Bernays did have a bit of a leg up when it came to the PR game. His uncle was Sigmund Freud, who was turning the world of psychology on its head with his theories of the power of the unconscious mind to shape our feelings, behaviors, and sense of self. As the world erupted into conflict once again and Hitler came to power, Freud began to think that there was some evil at the core of humanity seeping out. Inspired by his uncle, Bernays signed on to the notion that you don't buy a product just for its factual, useful attributes; rather, you buy because of an unconscious, emotional connection to the product. Frightened by how the

Nazi regime channeled what Freud and his disciples believed to be the innate darkness of human behavior in World War II, Bernays decided the best thing to do was to rein it all in by distracting us with things, like pretty dresses and dress slacks. In the words of Peter Straus, one of Bernays's employees, the strategy was this: "It's not that you think you need a piece of clothing, but that you will feel better if you have a piece of clothing." Shopping became a tool to control the masses.

Perhaps you or your friends and family can relate to this idea—I certainly can, and we know Claire can. Her social media addiction is a direct extension of Bernays's stratagem to turn people into distracted happiness machines. Buy first, and think and feel . . . never (or too long after the return date). Back then, the goal was to wave something new and shiny in people's faces so they wouldn't join something like the Nazi party—a (legitimate) threat of the day. Now, though, economic and political interests have been so entrenched in our fashion (ahem, Jeff Bezos) that we're distracting people not from some true evil, but from all levels of political engagement. Just think, when Claire's shopping, she's not doing or thinking about other things—like considering why Jeff Bezos earned the equivalent of the annual salary of his lowest-paid employees every eleven-odd seconds. (That is, until now, when she's reclaimed some of her time to consider those things and changed her shopping habits accordingly. Funny how fast that happened. . . .)

The same is true for a woman I'll call Tamara, a forty-five-year-old woman from Minnesota with bright eyes and great hair. Tamara's clothes serve two purposes, practical and emotional, which sometimes overlap: the former being clothes for her job as a hairstylist, where she stands on her feet for hours at a time and might get squirted with Lt. #27 and Orange Blonde Frost, and the latter being clothes that make her feel "cozy." During our hour-long conversation, she used the word "cozy" at least a dozen times, whether it came to her dress pants for work, or her "scrubby" sweats she put on when she was at home or running errands, or a blanket she scored at HomeGoods, where she shops even more frequently than T.J.Maxx. Clearly, emotions rule the day when it comes to her closet, and her wallet.

"I shop with my cash," she said. "Any money I make on the side—and I get tips—that money is my spending money. It gives me a couple hundred dollars a week to spend, like, foolishly. That's my wasteful money." But she wants to get the most for her "wasteful money" at discount stores like T.J.Maxx. Tamara stopped in at least once a week since it was conveniently located on her way home from the salon. She admitted that the bagful of stuff she walked out with left her less satisfied than the feeling that she's "always getting a deal" worthy of her tip money. Shopping at T.J.Maxx also meant that even a few hundred dollars of "wasteful money" got her more stuff. "I tend to buy things just because it's a good deal, not necessarily because I love it. Then I wind up not liking it at all anyway. I would like better stuff when I'm shopping. And when I get home it never feels like it's better stuff. Does that make any sense?" It did.

FROM CITIZEN TO CONSUMER: HOW FASHION STARTED THE MATERIALISM REVOLUTION AND SILENCED POLITICAL ENGAGEMENT

Retail therapy like what Tamara relied on is one thing. But scaled up, as Bernays wanted, shopping became a tool for social control. As violence threatened to tear apart civilization as we knew it and the economy needed the engine to keep revving, influential thinkers and political and economic elites in the United States picked up Bernays's work. Postwar, consumption became democratic, thanks to Bernays: buying jeans prevented social unrest through the distraction of shopping while also fueling the economy, à la Louis XIV.

In other words, shoppers like Tamara—who according to Freud and Bernays were plagued by unconscious desires that, if let loose, would cause evil and destruction—*had* to have those desires checked by the "rational"—mostly white, male—minds running the economy. People could buy their way to what they thought was happiness without disrupting civility and peace, and boost the GDP at the same time. They thought of it as a win-win. But for the—ahem—small fact that it meant replacing our primary role as citizens with that of our role as consumers, and corrupting the entire idea of democracy. In the words of Samuel Strauss, writing in 1924: "Something new has come to confront American democracy," which he coined "consumptionism," adding that "the American citizen's first importance to his country is no longer that of citizen but that of consumer."

Suddenly, the political system built for the people, by the people, was replaced with repression and compliance. Gone were the notions of participatory government, openness, and equality that Bernays had helped President Wilson share with the world. Historian of public relations Stuart Ewen described this new mode of democracy: "It's not that the people are in charge but that the people's desires are in charge. [They] exercise no decision making power within this environment. So democracy is reduced from something which assumes an active citizenry to the idea of the public as passive consumers driven primarily by instinctual or unconscious desires and that if you can in fact trigger those needs and desires you can get what you want from them."

Consumption had practical as well as ideological and psychological functions. After the war was over, this tactic served an immediate and utilitarian service. The United States had developed a robust war machine while fighting the evil Axis, turning out 185,000 aircraft, 120,000 tanks, and 55,000 antiaircraft guns in the wake of Pearl Harbor. The war perfected industrialization, from a production standpoint. Once evil was defeated, the economy itself became reliant on those factories continuing to produce, even if it meant turning out things like clothes and refrigerators (or chemical fertilizers and pesticides) instead of bullets. Millions of products could be produced continuously, which was far more than even the ration-weary nation could take in at once.

On the one hand, this was a good problem to have—and better than the opposite, too much demand and too little supply. Ultimately it led to the creation of a healthy middle class. But again, acting out of fear, businesses felt they needed to do something about the potential shadow side of their growth: possible overproduction. This scenario could only be alleviated by one thing: a steady stream of buying. In order to do so, as Paul Mazur of Lehman Brothers wrote, "We must shift America from a needs to a desires culture. People must be trained to desire, to want new things even before the old had been entirely consumed. We must shape a new mentality in America. Man's desires must overshadow his needs."

In other words, we needed to be *trained* to act upon the temporary feeling of joy we get from a great sale at T.J.Maxx, over the more rational thought that nothing in our cart would actually be useful to our lives and would instead, on balance, cause us more stress, as we will see. Tamara told me about the piles of workout clothes in her closets and drawers; when she goes to T.J.Maxx, she's usually looking for more athletic clothes. "I think if I buy the clothing, I will work out more," she told me. "I honestly have like fifty workout pants!"

Bernays was just the man for the job of shaping the "new mentality." Tasking himself with the mission of manufacturing desires at the pace of manufacturing stuff, Bernays married economic growth and social docility. His term for manipulating the minds of consumers was "the engineering of consent," that is, creating guaranteed conditions for people to say yes to a product, because its advertising had been tailor-made to satisfy their unconscious desires, just as we saw in *Mad Men*.

People today are exposed to up to ten thousand advertisements a day, whereas fifty years ago they were exposed to around five hundred. Exposure to those ads plays trickery with our brains by giving us a boost of the feel-good hormone dopamine, so we associate buying what's advertised with a kind of shopper's high. Coupled with the ease of access that online retail creates, you have the recipe for turning a whole population into civically disengaged stuff addicts, while the captains of industry accumulate vast wealth.

THIS IS YOUR BRAIN ON SHOPPING: THE NEUROLOGY OF ENGINEERED CONSENT AND A NATION BUILT ON CONSUMPTION

Thankfully, Freud was not the last psychologist to have his hand in marketing and its effects. Contemporary researchers are also looking into what happens to our brains when drunk on the cocktail of "engineered consent" and our contemporary distractions. Tim Kasser, professor emeritus at Knox College in Illinois, and author of a touchstone book on consumerism called *The High Price of Materialism*, has dedicated his career to the things that get us hooked on shopping, which we discussed one chilly February morning by phone.

It all started for him back in the late 1980s and '90s. Things were rosy on the surface—it was the era of free-economy-speak (aka neoliberalism) encouraged by Ronald Reagan and Margaret Thatcher. As we know, sending garment production overseas was one consequence, and the money was flowing from every part of the globe. But Kasser noticed something. He and his friends were all buying more stuff, but no one seemed any happier. The two conditions—unhappiness and extreme neoliberal capitalism— seemed related. Fast-forward to 1993, when Kasser and his graduate adviser Rich Ryan coauthored a groundbreaking paper whose name says it all: "The Dark Side of the American Dream: Correlates of Financial Success as a Central Life Aspiration."

His research has helped define five main correlates with materialism, which he defines as "a set of values and goals focused on wealth, possessions, image, and status." What he found was that people who demonstrate materialist values and goals also (1) have more financial problems;

(2) have lower levels of well-being, including lower levels of reported happiness, greater levels of mental illness, and more physical problems; (3) demonstrate more antisocial behaviors—for example, they're less likely to share; (4) show worse ecological attitudes and behaviors—which Kasser says is an even more robust correlation than well-being; and (5) have a suboptimal learning style, since if you just care about acing a test or getting into a school (i.e., seek status), you're less likely to actually try to learn the information and retain it instead of cramming it into your short-term memory.

Kasser believes that everyone harbors materialist tendencies. We do sometimes actually need things, and it's good evolutionarily speaking that satisfying needs gets rewarded in our brains. It's just that that reward circuit doesn't have to be turned on HIGH 24/7, contrary to Bernays's efforts. After years of research, Kasser has arrived at two main triggers of materialism: (1) social modeling, in which you do or believe something because of the example of someone around you—like your classmates, or the Instagram or TikTok influencer you ♥, or the millions of ads you come across as you go about your life; and (2) insecurity or threat, wherein you start to value goods as a way to satisfy a sense of protection, safety, or self-worth.

So what if our houses are crammed with stuff we don't truly want if it's not bothering anyone else? Well, first there is everything we just covered about the effects of materialism on how we behave in society. Then there's everything you've read in the previous chapters—the impact of how our clothes are made and delivered today on people and the planet. But if, even for the sake of argument, we exclude *all* of that, there is also a list of personal side effects longer than the one inside the Xanax box. Materialist values may have made it less likely to engage in evil politics, but in doing so, they also make for a whole lot of unhappiness. We have lost touch with the things that research has found would actually satisfy our needs *and* desires at a deeper level—connection, community, health, all of which get thrown by the wayside in a materialistic society.

Consider all the reasons one might buy jeans. There's the Bernays model: Someone I want to be like told me they're cool, so I want them. Then there's the human model: buying jeans with a sense of the welfare of the workers who made it, and if they make me feel cool or look good that's great, too. Or secret option number three is realizing that my real problem is needing to heal familial or social relationships, and jeans won't solve that. Those advertisers orchestrate the messages cooing softly from the blue light of our phones . . . *yes, these jeans will make you cool, and will donate $1 to a blind child in Africa.* So I buy the jeans, never thinking about the conditions of the workers being screamed at to go faster, or the chemical pool on the floor of the Chinese factory, or the fossil fuels burning away to spin those cotton fibers because, hey, a kid just got a dollar from me so my social karma is repaid.

Insecurity is an even more insidious trigger, which excessive materialism does nothing to alleviate. Insecurity can stem from emotional causes, sure, but there's also real, survival-type insecurity that is quite pertinent to the way fashion feeds off materialism. Let's say someone doesn't earn enough money to pay their bills—fairly common, as we have seen from the stories of garment and warehouse workers. Kasser's research suggests that those people will seek out *more* goods as a means of protection, and acquire those antisocial, antiecological, suboptimal learning qualities, too. Therefore, if we want to address the ways people engage with each other and their environment, we must also address people's financial precarity. Instead of fulfilling desire, we have to fulfill need.

Now, I'm not suggesting that all shopping should be a matter of life or death. I have derived real joy from shopping, and fashion has been meaningful to me and my relationship with my identity and now my work. But there's a way to root the two so that the pleasure centers being fired with our new jeans are *real*, not engineered by some man in a corner office. We need to reclaim our ability to exercise consent—in the jeans we buy, who we vote for, and how we live. Otherwise, we will have a world so warped by a false reality of docility and stability it can't help but implode on itself.

The good news is that we are now familiar with the larger system into which all our advertising now fits. Perhaps the pandemic has made us confront how much we engage with advertising from the comfort of our homes as it traps us in houses crammed with stuff, staring at screens that deliver ads even faster than Amazon Prime orders. We can choose to engage with that mindset, or not. Let's dive a bit deeper into how these materialist triggers have become our unwitting housemates and understand how to kick them out.

MODEL BEHAVIOR: CELEBRITIES, SOCIAL MEDIA, INFLUENCERS, AND YOUR BRAIN

The first of Kasser's triggers of materialism is one we all have succumbed to at some point or another: social modeling. It's basically the foundation of social media, the most conspicuous form of consumption there is. By broadcasting the appeal of stuff—jeans and sneakers, but also fancy vacation houses, beauty products, even the notion of popularity garnered by having the right stuff—and eating away at our attention, social media is what we consume and what consumes us.

But way back, when the internet was just a twinkle in Robert E. Kahn and Vint Cerf's eyes, Bernays began the trend of sparking social modeling–based consumption with celebrity advertisements. He was even hired by William Randolph Hearst, the magazine and newspaper mogul, to help use print advertisements featuring some of his celebrity clients like silent-film star Clara Bow (who had also hired Bernays for PR) to promote certain products. This extended to the big screen, where he launched the concept of getting products placed in movies and at film premieres (and marking a separation from official declared "advertisements" to ubiquitous, subconscious suggestion. Thanks, Uncle Sigmund).

These are marketing tools still very much in place today. Our favorite retailer Amazon is a perfect example. As we've seen, Amazon doesn't just sell toilet paper and books, or even just clothes, shoes, and accessories. It makes and sells movies, too. Bezos has justified Amazon's investment in Hollywood with the potential retail benefits, quipping, "When we win a Golden Globe, it helps us sell more shoes."

Over the decades, the role of the celebrity advertiser has only expanded. There are now many kinds. First, there's the OG celebrity—glam movie stars and singers/dancers whose popularity makes designers flock to them as free human advertising. Well, not exactly free. Whether walking a red carpet in a name-droppable gown or becoming the "face" of a brand, celebrities can bring in millions of dollars in income without having to memorize a line or a lyric. Compare these side hustle incomes to that of your lifestyle blog: Julia Roberts makes $50 million from Lancôme, Brad Pitt makes $7 million from Chanel, Blake Lively makes $4 million from Gucci, Cate Blanchett makes $10 million from Giorgio Armani, and Robert Pattinson makes $12 million from Dior. And the price tag of getting an A-list star to wear your dress? Somewhere between $100,000 and $250,000. That's more than one hundred times the annual salary of Rima, our Bangladeshi garment worker. For wearing a dress for one night.

Sounds like a dream gig, right? Who would bother doing "real work" when you could stand in front of a camera and have someone photoshop a logo on your photograph? Turns out that even the glam life of our celeb idols has a dark side. Because celebrities, whose stylized lives spark envy, if not joy, in the rest of us, are also victims of Bernays's materialist system. No one knows the struggle of making sure celebrities are wearing the right things better than the people who dress them: celebrity stylists. A profession within the red carpet industrial complex, the industry jargon that one insider clued me in on, stylists are essential to making all the moving parts of a celebrity's brand work—jewelry, shoes, and dress.

The celebrity stylists I spoke to, who asked to remain anonymous for the purpose of protecting themselves and their clientele, were frustrated by the way the women and men they work with are beholden to endorsement agreements at such high stakes. "They're essentially billboards," said one, noting how the long chain of handlers—publicists, agents, etc.—all have stock in something as seemingly simple as a dress and a pair of earrings. The red carpet is a branding opportunity that is significant for celebrities' careers. As another stylist told me, "They're selling themselves for future roles"; red-carpet attention translates to promotional might behind a film. And red-carpet appearances can also lead to advertising work, which as we now know can be extremely lucrative. The stylist continued, "There are a lot of financial and work opportunities for a celebrity that has a great red-carpet moment. So we are strategic because we want to provide as many future opportunities as possible."

A celebrity might get a huge paycheck for rocking a designer's look on the red carpet, but it's divvied up among the team first, which significantly reduces the number of zeroes. After everyone else gets their cut—agent, stylist, hair, and makeup—the "endorsement" that requires wearing an outfit, presenting a certain "face" to the world that might not be your own, doesn't pay as well as it might seem (comparatively speaking; it's still a lot of money). Plus, the whole situation is on a sliding scale of seasonal popularity. Brands will pay more for celebrities who are up for the highest awards in order to get more "impressions," as you might say, during the broadcast. That means the celebrity has to up their A-game on two fronts: win an Oscar/Grammy *and* look perfect.

We don't need studies to know that the ills of fashion disproportionately affect women—still true for celebrities. One stylist I spoke with mentioned that the conversation around women's bodies at award shows has started to change for the better since #MeToo, but there are other gendered factors at play when it comes to pressure to land high-paying endorsements. We now know that female actors tend to earn less for roles than their male costars. They often turn to these side-hustle ad campaigns, on and off the red carpet, to fill the pay gap. Again, the relative wage gap between genders of celebrities is in a different stratosphere from the gap between those celebrities and the people—mostly women—who work making and packing the clothes, as we discussed in the previous chapters—but it's a gap nonetheless.

Regardless of gender, the desperation behind the endorsement culture is only increasing as the wealth gap between haves (of all sorts—celebrities, financiers, rulers of nations) and have-nots increases. When placed in the company of noncelebrity billionaires, like at the Met Gala that Bezos cochaired, they feel like they have to suddenly top their own six- and seven-figure incomes. As Robert Frank, a wealth reporter for CNBC, explains, "there's only so much you can make in entertainment, so [celebrities] look around and decide that they need to get to the next level that they're encountering socially at the Met Ball and at charity functions." Wearing a dress one night is a relatively painless but quick boost to the top.

This mentality is partly why, until very recently (as recently as the 2020 Oscars, for which Elizabeth Banks wore a repeat dress in support of New Standard Institute), many celebrities have been loath to be seen in anything twice, even in tabloid-style shots of them running to the drugstore (because famous people need Kleenex and Band-Aids, too!). Each outfit was a free gift for free advertising, a required contractual plug, or an opportunity for one—maybe your jeans worn on a CVS run would be photographed, wind up on the cover of *In Touch*, cause a complete sellout, and get you a gig with the brand, so that you don't have to take another sappy rom-com role.

Social media has also exploded the marketplace for celebrities as human billboards, and it's not going away. Perhaps thanks to the pandemic, social media use grew 10 percent between July 2019 and July 2020, reversing the trend of decreasing social media use in 2018 and 2019. The opportunities for human billboarding at red-carpet events, drugstores, and in advertisements as brand ambassadors have skyrocketed from the mere millions of TV and magazine viewers to the billions of people scrolling on their phones all day long. Not only can celebs get more likes, which boosts their endorsement value, with a killer dress, but brands also get a boost that will help them sell their other products by face recognition.

Social media may be putting a fun-house mirror up to the true value and manifestation of talent for celebrities, but it's also changing the nature of celebrity itself and, by extension, who gets to market to us. Which brings us to our second group of celebrity advertisers—the twenty-first-century edition, comprising people skilled not necessarily in acting, singing, walking down runways, or being royals, but in telling us stories, stories that compel us to buy what they're wearing, or the makeup they're applying, or the vacations they're going on, the music they are listening to, or the couch they are surfing on. I am, of course, talking about influencers.

Here's how the influencer numbers stack up. If you've got somewhere between ten thousand and fifty thousand followers on social media, you would be a "micro-influencer," and you might get a few thousand dollars for a single social media post. Under one million followers, you're still looking at maybe $10,000 per post. But score more than one million followers, and your post profit increases by tenfold. YouTube content is especially valuable—hello, Beauty Vloggers—which can bring in tens of thousands of dollars per video plug post. Over on the newer video platform TikTok, influencer deals are exploding. Take, for example, its highest-earning star, twenty-year-old Gen Z favorite Addison Rae Easterling. She has more than 63 million followers, a makeup line, deals with Spotify, and now sells a lot of jeans as the face of American Eagle. All of this nets her an estimated $5 million a year.

It's easy to understand the business case for why influencers tend to wear new things. As is the case for celebrities, every post, story, or video is a billboard moment, regardless of whether they're being paid for it explicitly. And if you are getting paid to wear something, why give away two billboards when you were paid for only one? As Paris Hilton, arguably the original influencer, said, "Brands just send me things to wear because they want me to post it, I change outfits six times a day . . . I don't even like this stuff." But the unintended consequence has been to distort consumption norms for the rest of us. Even if we don't have a million followers, we start to feel the need for a new swimsuit for a beach weekend, a new dress for a family wedding, a new pair of shoes for a party. If the general population, too, cannot be seen in things more than once, we need a production cycle that's cooking with gas (and probably coal, too).

Brands have even been starting to value influencers more than traditional celebrities for this reason. To use the same marketing language, they "convert" better. To put that in nonmarketing language, it means that we trust influencers to guide our purchases, because, unlike the red-carpet folks, these influencers are "just like us," or, perhaps, our ideal versions of ourselves. Their audiences are also inherently more connected via social media—a literal application of the whole "social modeling" theory. According to McKinsey Reports, 86 percent of brands use influencer marketing, and their cinematic "storytelling" and marketing campaigns are gaining such important ground that they could make up 20 percent of China's online advertising market by 2023, or $166 billion. (Influencers also come at a comparatively lower price point than a top actor, so companies can recruit several influencers and reach millions of dedicated followers even within those huge ad budgets.)

The data backs up the idea that we place trust in influencers, especially for clothing: a quarter of all purchases among a survey of European shoppers were made from social recommendations, two thirds of which was a direct influence. Clothing is one of the top categories for influencer marketing, with 5 percent of the influencers bringing in 45 percent of the social-based purchases. Meet the new 1 percent.

Another important shift: Now, we can actually buy the single-use garments being marketed to us. Fast fashion has made putting cheaper versions of influencers' new looks on the rack in sync with their posts' industry standard, and in the process exponentially increased that two-season calendar of Louis XIV to fifty-two "micro seasons" a year. Sites like Boohoo and Fashion Nova let the TikTok generation buy what their favorite influencer wore yesterday for even less than H&M or Zara. Boohoo was growing at nearly 50 percent per year, until—big surprise—it was found to be paying its factory workers (who are mostly based in the UK) the equivalent of $4.40 an hour, which allows for the very quick design-production turnaround) and not providing sufficient protections against COVID-19. Despite the negative press, sales continued to grow by 45 percent during the pandemic.

MEAN GIRLS: SHOPPING AS SECURITY

And here's where the second of Tim Kasser's triggers of materialism enters the picture from a back door: insecurity and threat. This factor doesn't just affect those who are financially insecure, in terms of having their basic needs met. We, as a society, are being sold the idea that we are not good enough, that our everyday lives, even when we are stuck at home, don't measure up to the images that we see on our screens (check out #quarantinelife for proof). Whether it's making us pine after designer dresses, perfect mom jeans, or the coziest organic lounge pants, social media has become a 24/7 global stage for social pressure at every age. Its prescription for this sense of insecurity? More stuff, of course.

I met a woman I'll call Jessica at a café in Brooklyn one sunny morning in mid-February. It was unusually warm for the time of year, and her presence was bright enough to compete with the sun outside: in honor of the recent Valentine's Day holiday, she wore a magenta pink coat, burgundy scarf, and a maroon velour dress with silver embroidery and matching lipstick. A self-proclaimed trend junkie, she's a Rent the Runway (RTR) ambassador and was a member of the company's former "unlimited" membership (the company canceled the program in September 2020), for which she received eighteen to thirty new items of clothes per month. She has the style of a maximalist—"ostentatious" in her words—but the closet of a minimalist, since she only owns basics like sweatshirts and PJs.

Her membership allows her to get a new look for at least every day of her workweek, plus more for special occasions, without the baggage of, well, baggage. Living in a New York apartment means a small amount of space, especially when you're sharing with a significant other like she is.

Between bites of a warm pretzel, she made the case for having a new look every day. It let her try lots of new, trendy clothes that were *also* well made, as opposed to the fast fashion she had consumed, and later tossed, when she first moved to New York and traded her midwestern go-to outfit of jeans and a T-shirt for leopard midi-skirts. "Our society is very judgmental these days," she added, lowering her voice as if hoping no one could overhear her. "What if my friends and I are out and take a selfie? I can't be seen in the same outfit twice."

While Jessica herself was not as influencer-driven as many—she did follow a few skewed more toward high fashion—it's easy to see how influencers have driven the idea that clothes should be instantly gratifying *and* instantly disposable, lest you be judged. Thanks to the red-carpet industrial complex, and the curated "reality" of social media, we, too, believe that wearing an outfit once is the only way to get fashion cred (check out #OOTD). According to a Barnardos survey, one in three young women in the UK consider clothes "old" after wearing them once or twice, and one in seven consider it a fashion faux pas to be photographed in an outfit twice. In the UK, a survey by the credit card Barclay revealed that 9 percent of shoppers admitted to buying clothes on the internet for the purpose of posting on the internet. After making the post, they return the item.

Jessica's solution of using a rental service to indulge in the single-use garment is an innovative one, and may be more financially and environmentally conscious than buying and throwing away more disposable items. But what it's missing, along with the entire trend-based aesthetic of celebrity/influencer culture, is the sense of ownership in a bigger way: owning your own style. Constantly chasing trends, even through rentals, is not the kind of long-term solution we need to solve our industry-wide fashion problems. Does shipping a garment three times and dry cleaning it between each wear, all in the name of fun and Instagram followers, actually significantly reduce ecological impacts? The data on

RTR's ecological footprint is still not clear, but road transit and dry cleaning are both energy intensive. The environmental impact cannot be small.

Instead, what if we addressed our social media FOMO at the source? What if we could address the reasons we feel threatened about our self-worth rather than trying to buy ourselves "better" lives? Edward Bernays wasn't banking on our collective ability to do so, but I have faith, and some ideas, to prove him wrong. And it all starts with the little rectangle probably within a foot of your body right now.

WE NEED TO TALK: SOCIALLY DISTANCING FROM OUR PHONES

How did fashion rules for the red carpet infiltrate our day-to-day lives? How do you get all of the updates on your friends, family, sales from your favorite brands, people who swiped right on your dating app, new podcast episodes, emails from your boss, and breaking news? Your phone. Over 80 percent of American adults, and almost half of all adults worldwide, now own smartphones. Americans check their phones ninety-six times per day on average. More and more frequently, all those racing thumbs are not texting but shopping. In China, the average person spends at least two hours shopping online each day. That unbelievable number is only growing in the wake of COVID-19. Forty-three percent of participants in a McKinsey & Company study who didn't shop for clothes online before the pandemic now do, and almost 28 percent (more so among Gen Z and Millennials) say they're planning to not buy as much at physical stores in the future.

Spending more time on our devices shopping is harmful to our health in a few ways. Increasing consumption has unseen consequences: It destroys our resources, contributes to climate change, spikes the pace of work, and pushes garment and retail workers around the world to the brink of their physical and mental health, as we've seen in detail already. But shoppers, too, feel the pain of their worsening habits. Tech neck, back pain, poor sleep, and more physical ailments can all result from too much screen time, but what we consume on the screens also has psychological effects.

Constant exposure to misguided fashion messages on our phones works against a healthy relationship with clothing, and our bodies, in a few ways. First, the frequency with which we check our phones means we are getting more and more exposure to materialistic signals via advertising. It's one thing to inundate adults with such messages; ostensibly adults have higher amounts of self-control and self-awareness to make decisions about their exposure (though that is highly debatable). But younger minds don't have even that level of self-control, and kids are increasingly the ones exposed to ad-driven fashion culture. Studies have shown that children who are exposed to advertisement-filled television and smartphone videos and apps demonstrate more materialistic values. One even showed that kids as young as four and five years old were more prone to play with a toy alone versus with other children after they were exposed to ads for the toy.

There is a growing body of scientific evidence about the negative effects of cell phone use on our higher executive functions, impulse control, attention, learning and information retention, interpersonal relationships, and even our commitment to corporate social responsibility and our own workplace dynamics. We don't even need to be using the phone. Just having it nearby, or hearing or feeling the vibration in your pocket, can elicit these effects. One study even found that frequent media multitaskers had experienced decreased gray matter in the part of the brain that controls attention (among other things). No wonder Claire couldn't resist the alert coming from her Zara app, reminding her it was time to buy something—anything—new, crowding out any logical reasoning from her conscious mind.

Needless to say, more time online—including more time online shopping—is not making us happier. The effects of screen time are particularly startling in teenagers. A 2018 Pew Research Center survey found that 95 percent of American teenagers—more than adults—have smartphones, which they use for about nine hours a day. Some even sleep with their phones to make sure they don't miss an update. That's a huge exposure risk to all sorts of harmful messaging, and not just about fashion. The CDC has charted a disturbing trend in the rate of suicide among girls: the number of suicides steadily declined between 1993 and 2007, after which the numbers shot up. Now, correlation does not imply causation, but it is worth noting that Facebook opened to the public in September 2006, immediately preceding that spike in suicides.

And the amount of stuff we have is stifling us. Tamara's closet is a testament. She had crammed her clothes into two closets in her bedroom, originally his and hers but now just hers, supplemented by another closet in the office with her dresses and clothes she wears less often. Leggings lived in her dresser drawers, and sweats got a pull-out drawer under her bed. And it wasn't enough; she had been planning to tear down a wall to expand her bedroom storage space. When I asked how much of her clothes she actually wore, from all these different homes, the nervous laugh that came out before her words told me I wouldn't like the number. "This is a lot of honesty," she said, but "probably five to ten percent."

The evidence around the impact of stuff on our mental health is staggering. Darby Saxbe, a psychology professor at the University of Southern California, published an observational study in 2010 to measure cortisol levels among couples. She found that more than 75 percent of homes have so much clutter they use their garages exclusively for things, not cars. When researchers measured the couples' levels of cortisol, a stress hormone that is meant to peak in the morning and decline during the day, they found the wives—though not the husbands—experienced a less steep decrease in cortisol throughout the day. Their bodies weren't able to wind down naturally because of discussion about the presence of too much stuff, dampening their moods and increasing stress.

Saxbe told me that she saw our relationship to stuff as analogous to our desire for fast food. When it comes to the chemicals in our brain, the dopamine-pleasure spike we feel when biting into a greasy, salty sandwich, or a sugar-loaded cookie, is identical to what we feel when we see a good deal and add it to our cart (real or virtual). Our society has become more and more aware of smart food choices, ones that sustain us for the long term instead of filling an immediate, often emotion-driven, craving. So if we can eat in a way that prioritizes nourishing our bodies for the future, can't we also shop that way?

The fashion industry is certainly not the only factor behind the rapid rise of anxiety and depression, the latter predicted by the World Health Organization (in pre-COVID-19 times) to be the leading cause of the global burden of disease by 2030. But if celebrities and influencers of all stripes are today's role models, they should question whether their endorsement deal is worth perpetuating a cycle of materialism-driven mental and physical despair, distracting from our active engagement with society, the destruction of the planet, and the exploitation of its most vulnerable people. And we, the influenced, should question our role as well.

When your home starts to be overrun with clutter, especially unused or nearly new clothing, there's an attractive option: Get rid of it! Tamara could very well return to T.J.Maxx anything she decides is "horrifying" after an at-home try-on; but we know the litany of excuses she, and we, have for not doing so.

Instead, like many people, she resorts to donating her unwanted clothing, usually to the Veterans Association. We'll learn more about these donation centers in the next chapter, but the short version of a longer story is that only a tiny fraction of donated clothes actually gets resold to people in our neighborhoods. Because so much of our clothes these days are low-quality and intentionally disposable, they don't resell well, and wind up getting passed down a chain of organizations, and across continents. Ultimately, the large majority of the clothes *will* make their way to a landfill or incinerator, somewhere.

The same is true for returned merchandise. Although retailers try to hide what they do with returned goods, estimates by the French and German governments suggest that $689 million worth of consumer goods were destroyed in France in 2014, and $9.1 billion worth in Germany in 2010. Online shopping is making this practice more and more frequent among brands, since people have a tendency to buy multiple sizes of an item with the idea that they can ship back whatever doesn't fit for free. About 11 percent of all products purchased online are returned; clothing return rates are even higher. Sounds peachy, except that the brands don't want to invest in the labor that would be needed to inspect and repackage goods so they just get thrown away.

The more Tamara and I talked about her shopping habits and the way they actually make her feel, the less jazzed she seemed about her next trip to T.J.Maxx, or about her closet renovation. She still owned a lot of things that made her think, immediately or soon after purchasing, "You shouldn't have bought that." In fact, there was nothing, not one item, in all of the hundreds of items she owned, that she said she "absolutely loved." The happiness she sought out by shopping wasn't in any of the bags on her bedroom floor, the bins of sweaters at the top of her closet, or her color-coordinated drawers of tanks and tees. If happiness wasn't there, then where was it?

RECLAIMING OUR VOICES IN FASHION: HOW TO SPARK JOY, AND THEN SOME, WITH WHAT WE WEAR

One of the first, and most successful, advertising campaigns modeled on Edward Bernays's theories of engineered consent was also one of advertising's biggest coups. The campaign was for cigarettes. (If you've watched

Mad Men, you'll recall how big a deal the Lucky Strike campaign was for Don Draper and his crew.) For several decades in the early twentieth century, major brands like Lucky Strike, Camel, and Virginia Slims used a host of creative means to encourage people to light up. Including one that equated women's empowerment with smoking that was led by Bernays. As innocent people flipped through magazines or drove their cars down the highway, images of smiling medical doctors, movie stars (including a 1950 ad for Chesterfields featuring a young Ronald Reagan), athletes, and even animated wildlife flooded their retinas, telling them it was not only good, cool, and liberating to smoke, but safe—even using ad copy to assert that doctors recommended their brand over competitors. The brands were happy, and the smokers were hooked.

Then came January 11, 1964. The US surgeon general issued a report (on a Saturday, to maximize media coverage and minimize the impact on the stock market) naming the many health detriments of nicotine and tobacco, including lung cancer, chronic bronchitis, birth defects, and a 70 percent increased mortality rate. By the following year, 1965, Congress had passed a law requiring all cartons to bear a health warning, the now-commonplace "Surgeon General's Warning," which was meant to give customers pause before they bought their next pack. Within five years, all TV and radio ads for cigarettes were banned. While the industry still spends billions of dollars a year advertising their cancer sticks, the bans on advertising on TV and radio clearly correlate to declines in smoking. In the fifty years since the original surgeon general's report, which spurred the anti-ad legislation, America's smoking rate has been cut in half.

What does this have to do with clothes? Sure, both smoking and shopping (especially online shopping) are habit-forming, even addicting, not good for your health, and bad for the environment. But I'm not here to tell you not to smoke (though, I mean, you shouldn't). Instead, this example shows what can happen when public knowledge and legislation unite. We'll get more into legislation in chapter 9, but for now, and in light of this chapter's revelations about what your mind experiences while shopping, what I'll tell you is this: We have brains that are ours to own and act with freely. We have the cognitive control to seek knowledge about what we wear, *and* derive joy from, whether it's clothing or not. And when we use our brains, and our voices, we can also speak out for the kinds of changes that helped alter America's nicotine habits, and can help us change our current, trend-centric, influencer-fueled, disposable-fashion habits.

And we can start in our closets—or, really, pre-closet. Namely, by reclaiming the subconscious, innate, personal preferences and emotions that marketers like Bernays try to manipulate, we can make a stand against poor shopping decisions and reprioritize our identities as citizens. You can reclaim your personal style, and create a closet that frees you from the cycle of consumption so you can wear your clothes to live your life, not live to wear your clothes. In the words of astute fashion journalist Rachel Tashjian, "Personal style, not fashion, holds the greatest reward: it allows you to invest in *yourself,* rather than in a bunch of ideas about who you could or should want to be."

We can look to the people in this chapter for some clues about how to hone a personal style, if that whole concept is leaving you short of breath or scratching your head. First, take a look at the clothes you wear the most. How are they made and what are they made of? Claire didn't miss an enthused beat when I asked her what her favorite materials were. "Cashmere and silk—it's soft and comfortable," she said. These pieces are in line with the vision of her ideal wardrobe, one that's made of "beautiful pieces in great materials." She emphasized a key difference between French and American style, namely that Frenchwomen dress their bodies; they don't just follow trends and try to make things work that won't. (That, and what she referred to as "the Dyson," or the windswept look of American women sporting blowouts on the streets of Paris during Fashion Week.)

It's true that when I lived in Paris with Claire, I saw my own style evolve to a simpler, more timeless, more thoughtful and curated, and sometimes less-blown-out look (though I admit I do own that Dyson hair dryer). It's something I recognized in the American shoppers I talked to for this book, too, who either couldn't identify their favorite fabric (Tamara assumed hers was a synthetic, for its "coziness," until she looked at her favorite shirts and found they were cotton), or knew their preference but had less ownership over the material or understanding of how cycling it through a dry cleaning ring might not be as "guilt-free" as they thought. What this tells us is that more shoppers need to know what they're actually buying and *own* it—not just by hanging in their closets, but by making the decision to buy, and making its place in their homes and on their bodies part of their personal, less-influenced, brand.

It's not all our fault that we don't know what kind of clothes we are wearing. Bernays, advertisers, and the fashion industry designed the system to be thoughtless: see ad, buy, the end. (The same mind-numbing is baked right into the production and distribution of clothing creation.) The speed of our disposable fashion cycles doesn't help either. But educated shopping is at the core of what Paco Underhill, retail anthropologist and author of *Why We Buy,* predicts will lead us into the future of fashion. "Educating people at an earlier and earlier age that there are very few things that are transformative," he told me, is going to check people before they buy, resulting in fewer regretted purchases that get returned, lobbed in the back of a closet, worn reluctantly, or kept and then tossed or donated. When people let go of the idea that one more thing will make them somehow different or better, they get clearer on who they really are, what they want, and what clothes can support that. And once those items are found, they'll be more interested in caring for them.

Finding out who "you" are style-wise means pushing against every grain of marketing know-how of the Mad Men dynasty, including the idea of market "segmentation," in which marketers compile data about shoppers' interests, demographics, and other likes to customize ads that make them feel like their unique needs are being met, and that the items they're being sold are actually going to enhance their lives.

Identify what's really us (a person with a complex and evolving sense of self) versus what box advertisers want to put us in (a "woke" woman who wears mom jeans). Be skeptical of what marketers try to tell you about yourself. Consider the idea of "spiritual consumerism," wherein brands are infusing their products with feel-good messaging about helping a social cause, improving your health, or literally cleansing your soul with a tri-folded T-shirt à la KonMari, all in the name of a purchase. If we want to be more spiritual, we can just do that, and, better yet, it's free!

Or consider "marketplace feminism," the idea that we are supporting feminism with our wallets. A classic example is how the (excellent) manifesto *We Should All Be Feminists* by Chimamanda Ngozi Adichie became trendy; for example, Dior splashed the title across a T-shirt that cost $860. I'm not aware of Adichie getting a cut—if so good for her—and I can't imagine she would tell a reader of hers to use that amount of money to buy a shirt instead of supporting a feminist nonprofit working to take down legal barriers to progress, or just keeping the money and living a good life. When we think we can buy our way to a better society, all we do is create a society where buying is the only way we matter.

So, back to our original question, can shopping make us happy? Yes, but first we have to know what happiness means, looks like, and feels like to us—not what Don Draper says it should. Maybe it's a luxe cashmere sweater, or maybe it's a pair of secondhand jeans with patches that you wear to a phone bank to make calls for your favorite political candidate. Either way (and in all the thousands of shades of gray in between), you'll know you have made a sound sartorial choice when you feel fully in control of your purchase every step of the way—that is, fully human.

Seven

Tidying Up: What Happens to Clothes When We Get Rid of Them

One August day I found myself on top of one of the most impressive mountains I've seen with my own eyes. I'd scaled it with very little equipment, in the same Nike sneakers I wear to run on the treadmill at my local gym in Williamsburg. There were no harnesses, no guides, and no resupply stations along the way. In fact, it only took me about twenty minutes to get to the summit, despite frequent picture breaks. At the top, I looked out into an endless blue sky to my right, and a cloud of furious black smoke to my left.

It was not your typical mountain. Located about twenty-five miles away from the capital city of Accra, Ghana, the Kpone landfill holds household trash, plastic bags, food waste (though surprisingly little that I could smell), and—you guessed it—clothing. Although the majority of the actual trash there was occluded by the plastic garbage bags it was collected in, I hardly needed to bend over, let alone pick through, anything in the giant garbage heap I'd climbed to find dozens of recognizable labels: Puma high-tops, H&M shifts, knockoff Versace bags, even other Nikes like the ones I was wearing. Oh, and as it so happened, that day a portion of Kpone had caught fire—hence the black smoke. Over the course of the week, the entire landfill would go up in flames.

If you're wondering how a lot of foreign-label shoes, clothes, and

The Kpone landfill, with ominous smoke clouds that had already engulfed the other side.

accessories arrived in a landfill in Ghana, you're not alone. Very few of us think about what happens to our clothes after we've gotten rid of them. We lug garbage bags full of items that no longer "spark joy" to the Salvation Army, or we toss them in the trash with our Starbucks cup. We donate with good intentions—we want our things to have a second life, and believe they do; we want someone else to get good use out of them, even if we've decided they're useless. And while that sometimes happens, the truth is that there is not enough global demand for the massive quantities of secondhand, low-quality clothing we donate. As a result, our good intentions become costly, overwhelming waste and an environmental nightmare for people living halfway around the world.

Remember the statistic I shared in the last chapter about how the square footage Americans take up has almost doubled since the 1970s? Despite all this extra space, we throw out more than eighty pounds of textiles *per person* per year. Of the 267.8 million tons of trash generated by all Americans in 2017 (which averages out to 4.51 pounds per person, per day), 4.8 percent of it was clothing and shoes. That's almost 12.8 million tons.

And five minutes later, the same landfill as the smoke from the fire increased.

As we'll see throughout this chapter and the next, and per the somewhat oversimplified chart on page 182, discarded clothing goes on a journey unto itself, one that's just as complicated when it comes to human and natural resources as the production/manufacturing journey we followed in the earlier chapters of this book. Once we decide to "get rid of" a garment, it becomes several other people's problems, until, and often far sooner than we think, it winds up as trash. And just as the raw materials of our garments traversed tens of thousands of miles to make their way to us as the garment we may have been cajoled into buying via Instagram, and may never even wear, the end of its life may demand just as much travel, and may pass through just as many hands.

The ease with which we dispose of our garments is very much part of the problem we're discussing in this book. This chapter takes us through the part of the process that happens domestically, bringing to light the reality most of us have been hidden from thanks to well-equipped, functioning sanitation systems and a donation system that seems to itself not fully understand the unintended consequences of its work overseas. In the next chapter, we see the full impact of our domestic systems on the rest of the world, returning to Kpone among other places (get excited). Below is a visual of where we'll go in these next two sections, a path that reveals how while our trash can sometimes be another person's treasure—mostly it's just more trash.

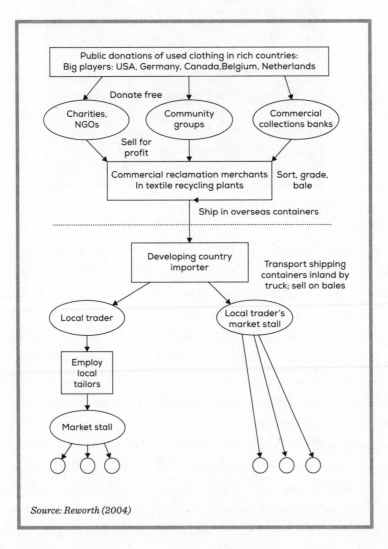

Source: Reworth (2004)

DUMPSTER DIVE:
GARBAGE COLLECTION IN NYC

"Nothing's terrible, nothing's impossible," said Vito, the twenty-seven-year-old man who was sitting in front of me on a cold day in early 2020. I'd seen jeans removed from the plant as cotton fiber, spun, dyed, woven, cut, sewn, washed, finished, packed, packaged at Amazon, and sold in the chaotic online marketplace inside our phones. I was ready for the average person's next step: throwing them in the garbage along with banana peels, Amazon bubble envelope, and the other 200,000 tons of clothing and textiles New Yorkers toss out each year. Vito, a New York sanitation worker, was explaining what would happen if I threw away my jeans in my home in Brooklyn.

Vito greeted me with a firm handshake and a warm, kind smile at an upscale café near where I work in Williamsburg. He was wearing just a hoodie, despite the frigid temperatures, which made the message of his stocky frame and tree-trunk-like, tattooed forearms even clearer: "I will have no problem picking up your trash." Still, his soft-spokenness caught me off guard more than a few times. He repeatedly declined my offers to buy him coffee, and only at the end of our talk did he explain why: "I'm more of a Dunkin' man."

With a Staten Island accent that sounded straight out of central casting for the next Martin Scorsese mob saga, Vito explained to me just how he got into this line of work. A few years ago, he took every civil service test required for work in many city departments—medical, fire, police, and sanitation. He got a call from the City of New York Department of Sanitation (DSNY) less than a year after passing the test, a chance so rare he jumped on it; the position is so coveted one could wait four or five years for an opening.

It's a "great job, city job," he said, and the benefits are good, too. It does mean that his sleeping habits may not be ideal, because it requires odd shift hours at six a.m. to two p.m. or four p.m. to twelve a.m., but between the protections he gets from the union and overtime pay (which definitely ramps up when it gets snowy), he doesn't mind. Plus, his great-uncle rode the trucks for twenty years, so it feels a bit like the family business to him. I was frankly surprised by how positively he spoke about this objectively difficult, literally smelly and dirty—but union-supported—job, especially in comparison with the universal dissatisfaction I noticed from Amazon workers. It may seem like Vito has the worse (and smellier) deal, but his seemingly genuine satisfaction with this work made me think Laura, Sam, and Becky might swap their step-ladders and tape rolls for union-backed Hefty bags if given the chance.

The longer I talked with Vito, the more I realized just how many things about garbage are nothing like I'd imagined. The streets of Williamsburg, Brooklyn, that I call home are far from overrun with trash, but that certainly does not make New York, or the rest of the Western world, trash-free. All that waste does not disappear on its own: It costs approximately $2.3 billion a year to get rid of, paid for with our taxes and commercial hauling costs. Though it costs a pretty penny to remove and manage garbage, like the other unsavory aspects of our buying habits, trash in NYC, and for most Americans, is out of sight, out of mind. Workers like Vito pick it up mostly at night or in the early morning, so we enjoy really clean sidewalks during the day without much disturbance to our daily lives.

Even with that relative naïveté about the city's well-oiled trash machine, I knew that my neighborhood had certain days for different kinds of trash collection, like regular garbage versus recycling. When DSNY employees arrive for work in the morning, they don't necessarily know what they'll be assigned to, so it's anyone's guess where they'll be and what they'll encounter.

"What [day] is considered the best?" I asked Vito, wondering how the hierarchy of trash would line up for this expert.

"Uh, my personal opinion? I would say garbage collection."

"Why's that?"

"It's . . ." He started to say "easier," but stopped midstream to clarify. "Well, it depends where you are. Some areas have cleaner garbage, some areas don't care. It's all mixed. You have people who come by who slice open the bags and try to look through them, and when you go to lift them and turn them upside down, it all comes out of the slits." Who knew there was such a thing as "cleaner" and "dirtier" garbage?

Although Vito claims he's "never [encountered] anything where I'm like, 'uh, this is terrible,'" he did share that recycling day was lower on his list of "commodities," or types of trash. "When you're on recycling," he said, "your day is a little bit longer. Garbage will take up more room on your truck. Recycling, it's thinner and smaller. Glass and plastic cracks and breaks." More recycling fits in the truck, so he might finish his route but have to stay out longer in order to fill up to capacity. Vito doesn't gripe about recycling his own trash. "It's like jury duty," he said, surprisingly positively, a civil service we all need to do in order for the system to work.

But for most of us, it seems, recycling is like jury duty in an annoying way—perhaps why we're not great at it. A full 56 percent of the city's recyclable waste winds up in landfills despite there being people like Vito to take it to the proper facility for recycling, if we just put it in the right container. At this rate, Mayor Bill de Blasio's goal of eliminating 90 percent of waste sent to landfills from New York by 2030 seems rather lofty, especially considering $106 million in budget cuts to DSNY thanks to pandemic shortfalls. At the time of writing in October 2020, compost and electronic waste pickup was suspended, and trash has been piling up in street corner wastebaskets because of reduced pickups. Sunday basket trash pickup was suspended, and people are throwing more and more stuff away because they're stuck at home—with all that stuff.

Changes to waste removal during COVID-19 have caused public health concerns not just for those who choose to brave the streets. Sanitation workers are susceptible to contracting and spreading the virus while doing their jobs. Social distancing is challenging for workers; garbage is a two-person job, and staying six feet apart inside a truck is difficult. Wearing a mask and other protective equipment while on the route sounds pretty darn uncomfortable. Then there's the issue of exposure to potentially contaminated waste. The daily garbage route became one of the front lines of the virus; sanitation workers are frontline heroes who provided a crucial civil and social service to our communities.

After our discarded jeans make their way onto one of the 2,200 garbage trucks cruising around town, maybe manned by Vito, they'd be dropped off, along with your neighbors' takeout containers and old Ikea furniture, at a waste transfer station, a local central depot of trash, before they make their way to their final destination. This is where I met Chief Keith Mellis, who oversees garbage collection for all five boroughs.

Chief Mellis pulled up to the curb to pick me up after I'd made a rather harrowing bolt across the crosswalk-less street under the Gowanus Expressway to get to the Hamilton Avenue Marine Transfer Station. I'd driven along this road many times in taxis, and yet somehow I'd never even taken a second glance at the building where all of the waste for much of Southern Brooklyn stops on its journey to a landfill in Virginia. Chief Mellis explained to me that the very modern waste transfer station I visited in Brooklyn was designed to be invisible, and integrate right into the community. Built in September 2017, this station can process up to 1,600 tons of garbage per day—to get the picture, imagine 1,600 adult brown bears walking through the city. The entire city, between domestic and commercial waste, produces around 24,000 tons per day.

The trucks are driven up a ramp where they are weighed and checked for hazardous waste; we had a special pass to go through in Chief Mellis's car (which, to be clear, was *not* a garbage truck). We then pulled into a vaulted concrete space. In front of us was a very large pit where the trucks dumped their loot. The dividing line between the parking spaces and the pit was the "tipping stage." What happens there probably isn't a surprise: The garbage trucks pull up and tip their collection down into the pit. While the chief made small talk with Chris, the superintendent of this facility, I watched with mouth agape as a stream of white and black garbage bags, a few mattresses, and a healthy flow of garbage juice (called "swill") came tumbling out of the back of the trucks that came rolling in like a motorcade. The clock had just struck nine a.m., peak time here at Hamilton.

What happens to our trash from the platform is a dance of industrial engineering. First, like a real-life video game, all those garbage bags and large debris are continuously transferred by enormous wheel dozers into four deeper pits toward the back of the platform. The whole process looked positively Sisyphean. As soon as the platform was cleared from one garbage truck dump, a new truck would arrive to empty out its contents with an echoing crash. Digital scoreboards mounted on the back wall of the space keep track of how much trash is being compacted into each pit. As the wheel dozers pushed the trash to the deeper pit, another vehicle with a large weighted boom came over to smoosh all the trash, making it as compact as possible in order to maximize how much trash it held.

"How much exactly fits in there?" I asked Chris.

"Eh, fifty thousand pounds," he said without blinking. I imagined the workers there playing some kind of game to see which scoreboard reached 50K first, arcade bells dinging whenever a pit got full.

After several minutes mesmerized by the tango between the vehicles on the floor, Chief Mellis and Chris led me inside to the command station. Standing beside two very chilled-out operators, I peered through a window to watch what happened to all of that compacted trash. The pits that I saw on the platform were actually emptying into below-ground containers. From the control room, I could see both the trash floor and the Gowanus Canal outside, where those containers stood like an army. Our jeans would likely have arrived in the country by container ship and now they would leave the state in another massive container, after just that brief stop in our closets.

A swarm of seagulls flitted around the containers, hunting for something to snack on. But these airtight vessels didn't allow much smell or any spillage to come out during transport. This was a big improvement from the open-air barge that used to bring the trash to Staten Island, Chris commented, back when the city still had the capacity to landfill its own trash. Indeed, I was surprised by just how little smell there was in the whole place, thanks to an advanced dust collection system that keeps smelly particles at bay.

Like Vito, the crew at the transfer station is pretty intent on having "clean garbage." Legally they need to demonstrate cleanliness by having the floor completely clean for one hour a day, a health requirement. Some private centers choose to take their hour at night when it's not so busy. Here, though, Chris's team takes it up a notch. They're always cleaning, he explained, hosing down the floor with water whenever they can. Here it was, the nine a.m. rush, and half the floor would be full of garbage by the time all the trucks came to dump their morning shift's collection; but by noon, the whole floor would be spotless.

New York City takes its garbage seriously, something I didn't quite appreciate before visiting the Hamilton Avenue Marine Transfer Station. Chief Mellis and Chris were both experts in the process of trash collection, taking pride in how their work makes the city clean and orderly—values reflected in their military-inspired uniforms and titles. And in a city of more than 8 million people, trash is no joking matter. They're responsible for making sure the 3 million tons of waste that the city exported in 2019 (each household produced 46.1 pounds of textile trash alone) gets into those containers seamlessly. The whole dance was so well choreographed, they kept having to usher me along when I got caught up staring at the proceedings. Hitches in the choreography are rare, and quickly and expertly fixed. For instance, while I was watching one container getting "lidded," an alarm started to scream out the signal that something wasn't right. A sanitation worker in a green jumpsuit strode over along the narrow metal catwalk, politely asking me to move out of his way, and with a flick of a wrist adjusted a screw that hadn't caught properly because of the cold.

Like Vito, none of the men I met with at DSNY—there are female employees; I just didn't see any that day—in any position throughout the process, seemed unhappy with their work. Every job has its own form of garbage—whether it's literal or metaphorical—and yet this one actually came with some perks that made them all seem to step up to the task with pride, even in summertime—peak trash season—when the city hosts its street fairs and tourists. Twice a day, they empty the green sidewalk bins filled with summer-fun accessories like picnic paper plates and table-cloths and single-use ponchos for those July thunderstorms. They did a good and critical job for the city. As Chief Mellis said, "There's not much that New York's Strongest"—that's the term for New York sanitation workers—"doesn't do."

Final Destination: Clothes in Landfills

But all good things come to an end, including DSNY's fastidious reign over the trash. They wash their hands of it all as soon as that lid is sealed onto the container, when the trash becomes the responsibility of Waste Management, a massive public company. It ranks #207 in the *Fortune* 500 list of largest companies as of May 2020. Though New York City doesn't outsource home garbage collection jobs of people like Vito, it has out-sourced the garbage itself. New York sends its garbage to landfills and in-cinerators (more on that soon) far and wide; the closest is in Fairport, New York, way up north near Rochester, about 250 miles away, but there are others in New Jersey, Pennsylvania, Virginia, and South Carolina. In 2019, the city spent $411 million plus another $87 million on disposal costs to export this trash out of its already teeming streets, up from $300 million in 2014. The cost to send away trash is projected to increase to $420 million in 2021. That comes to $512 per ton for *just* refuse collection, and $202 per ton for disposal. And when we're producing more than 3 million tons of garbage per year in the city, per 2018 records, that's no small bill.

Waste Management gets the trash out of the city: first by barge to Elizabeth, New Jersey (on its way, it waves hello to the Statue of Liberty), then by railcar 384 miles south down the coast, to Atlantic Waste's 1,300-acre landfill (a unit of Waste Management) in Waverly, Virginia. This process helps to reduce the road traffic in the city, and thus the general pollutants for New Yorkers in all the boroughs. The city has a twenty-year contract with Waste Management, which itself has permits with the local government of Waverly to allow *our* trash to be *their* neighbor. Keep in mind, though, that the location of any landfill is not just about where there is space to dump garbage. Race correlates to one's proximity to toxic waste, as noted in chapter 6; how close you are to hazardous trash depends on layers of social and economic prejudice.

It might be obvious why being near a landfill would not be something to call out on a real estate listing. There's the smell, the visual, and the fact that, as I learned from trash and recycling expert Peter Anderson, lecturer on recycling systems at the University of Wisconsin, landfills are basically like "very large diapers." In an animated, raspy voice that often sounded out of breath (with exasperation, I gathered), Peter explained to me by phone what exactly goes down at a landfill. Once a site is selected for this special honor, a barrier is placed on the ground so that liquids from the trash—the "swill" I saw in Brooklyn—do not contaminate the surrounding environment, and particularly the groundwater. Piping is put in place to remove this liquid, called "leachate" (one man's "swill" is another man's "leachate"), which is then treated in a separate location. New loads of trash are compacted repeatedly to remove as much of the air as possible, making a landfill essentially anaerobic. At certain intervals, a compacted layer of soil or other cover material is added to minimize odors, pests, and rodents. Layer by layer the landfill grows, just like the heap of unwanted clothes in Tamara's closet. When it becomes full, typically within thirty to fifty years, the lifespan of the modern landfill, it is sealed with clay, plastic, or soil.

So here are our jeans, having been transported and compacted in

Brooklyn, and compacted again down in the Virginia landfill. What happens next? All clothes are not created equal, as we well know at this stage. And that inequality carries over even when we toss them out of our closets. You see, organic products—organic with a lowercase "o," meaning derived from living matter like cotton, or more commonly food scraps— will break down over time when exposed to air and moisture. Depending on where your jeans are in the landfill, they may or may not break down; if they do, that's not a good thing, as they'll produce methane gas, a powerful climate-change-causing greenhouse gas.

Then there are synthetics, which are not organic material and thus won't break down under any landfill conditions. Think about all those jeans with 5 percent elastane, or other blends of natural and synthetic fibers. When they are tossed in a landfill, the organic part might break down, producing climate-change-causing methane, but the rest, made of fossil fuels, would not. Those fibers that helped make your butt look good for the handful of times you wore them just stick around, and, according to Anderson, "the stuff there is dangerous for a thousand years or more."

Here in the United States, there are other ways to dispose of clothing, too, though they may be less common. Trash from the West Side of Manhattan, for example, goes to the Covanta waste-to-energy facility in Newark, New Jersey, which has a "clean" incinerator; nationwide 12.7 percent of our trash, or 34 million tons per year, is burned. While the term sounds nifty, and is certainly superior to open burning (which I describe later) because it captures some energy for the grid and filters some of the pollutants, it is far from a perfect process. In addition to emitting climate-change-causing CO_2, these facilities have also been found to emit pollutants harmful to health, including particulate matter, dioxins, lead, and mercury. The people who are exposed to these pollutants are most often communities whose residents are predominantly minorities or low-income; the Newark facility is in a community that is 71 percent minority, in which 37 percent live below the poverty line. All this is to say, it is not a good look for the West Side of Manhattan, which is home to some of the wealthiest zip codes in America.

The politics of trash are as tense as the rest of our political system. Once upon a time, back in the 1970s, when waste was being dumped in rivers, oceans, and open landfills, citizen campaigning (like what drove the creation of the EPA) resulted in laws here in the United States outlining how landfills must be built and maintained to prevent hazardous materials from being tossed out with your takeout containers and jeans, and to prevent hazardous liquids from entering the groundwater. But those regulations are still too few, and rarely kept up after a landfill is filled and sealed up. After thirty years, for instance, no one is responsible for what happens to a landfill—even though the decomposition process takes decades, even centuries, more. The government metaphorically washes its hands of that waste, leaving it to local communities to worry about. And, again, it's critical to note how structural racism finds its way into our trash, as race is the biggest factor that determines who lives in these communities next to landfills.

With its goal of becoming almost zero waste, to avoid these outcomes and ostensibly to help the planet, New York City has promoted clothing donation to divert textiles from its waste stream. The city has a number of ways that clothes can avoid an encounter with Vito, including weekly clothing collections at greenmarkets and periodic "Stop 'N' Swap" events, for people to exchange clothes and housewares and walk away with something "new." Since 2007, 6.25 million pounds of clothes have been collected at these offerings, and yet 6 percent of the city's waste stream per year continues to comprise clothing, linen, shoes, and accessories.

The national stats for reuse and recycling of clothes aren't much better: It is estimated by the EPA that only 16.2 percent of textiles (which includes clothing) are "recycled." However, as we will see, it seems this statistic is an aspiration rather than a reality, as much of what is included as recycled actually becomes part of a massive international secondhand industry, and much of that does *not* in fact get recycled.

TRASH TO TREASURE:
WHAT HAPPENS TO OUR DONATED GOODS

If throwing your jeans in the garbage bin was the first option for ridding yourself of a garment, option two for most people is donation. For better or worse, the influx of donated goods is on the rise—doubled, in fact, over the past fifteen years—for a couple of interesting reasons. First, there is the Marie Kondo effect. The books and Netflix show from our favorite Japanese tidy-er has sparked in people all over the world the desire to purge. And purge. And purge again. Donation centers were overwhelmed when the show launched in January 2019. The pandemic only accelerated this change; more people were spending more time at home, and closet cleaning became a source of sparking joy. ThredUp, an online secondhand clothing retailer, found that garment disposal was up by 50 percent from pre-COVID times. The site reports six times more requests for its mail-in donation kits. Meanwhile, bags of clothes left outside closed in-person donation centers will likely just be sent to a landfill rather than properly sorted and sold.

While donors likely believed their clothes would go on to spark joy in someone else, or perhaps clothe a homeless person, only about 20 percent is actually passed on to a person in the way we might imagine. In other words, 80 percent of donations leave a donation store's floor not in the arms of individuals, whether they're needy or people who just like "thrift-ing," but packed into huge bales headed many places, including overseas.

One chilly January morning, I trekked to the Salvation Army pro-cessing facility in midtown Manhattan to meet "Major" Fred Muhs, the center's director, who would give me the inside scoop on the donation reality. Major Fred greeted me with a firm handshake and bright smile. His second-floor office had a 1970s vibe, which made sense once he told me he'd been "serving" in the Salvation Army for thirty-four years. His wife, son, and daughter also work for the organization, so the Salvation

Army is a true family calling for him. The Salvation Army uses similar military rankings as the real army, and Major Fred has the uniform—button-down shirt with red shoulder signifying his rank.

Founded in the late 1800s in England, the Salvation Army started off in the collections industry with a fleet of pushcarts that wandered around cities collecting unwanted paper. Eventually, as the country became richer, people started wanting to get rid of their clothes, too, not because they could no longer be used, but because they were no longer wanted, so the paper guys would throw them in their cart and bring them to HQ. Not knowing what exactly to do with people's tattered old dresses and suits, the collectors tore up most of those early donations and sold them to blacksmiths and other tradesmen. As time passed, and more and more clothing started to come through their doors, they figured out that the garments could help fund their charitable efforts.

The profits made from the sale of the garments at the location I visited fund a rehabilitation center a few blocks north, providing room for 125 people. The Salvation Army's services have little to do with the clothes it collects; guests don't receive free clothes at the centers. Instead, our discarded, donated clothes help fund the services of warm beds, food, and other necessities for the needy.

Major Fred's facility receives a staggering number of pieces—up to sixteen thousand pounds of clothing a day, or nineteen thousand garments. But that's not all the Salvation Army collects in the city. Another facility in the Bronx receives another ten thousand pieces per day. Fred's best month of the year is October, with March and April coming in next. He theorizes that it's because that's when the seasons change. People are clearing out their closets and wanting new styles they can wear now, affecting both donations and purchases. Donations to the Salvation Army overall peak in December, specifically the last week, in order for us to sneak in one last donation item on our income taxes.

The average garment is sold for $3.99, with jeans priced slightly higher at $6.99. A little mental math puts us at about $76,000 a day, if everything received is sold. Not chump change for clothes that were donated. For free. In 2017, one of the Salvation Army's main competitors, Goodwill, generated $5.9 billion in retail sales; the entire thrift industry brought in $17.5 billion in revenue.

Looking around the bustling center, it's hard to imagine the humble beginnings of what Fred calls "the nation's first recyclers." Back then, shoppers would come in and rummage—hence "rummage sale"—through bins of clothing to find the perfect thing they didn't know they wanted, or, more likely, simply something to put on their backs. Today, things are a little different.

The bin system of yesteryear has gotten a few upgrades. Remember that Fred's center isn't just getting neighborhood folks passing by with a garment or two. He's getting tens of thousands of pieces every day, and not all of them can, or should, be sold. When I asked him how he determines the donations' fate, he pointed upward, as if toward the heavens, to the fourth floor: the sorting floor. This is where employees, appropriately called "sorters," pick through, by hand, all of the clothes that come in and determine what is saleable. The requirement (which is entirely subjective) is this: Anything they would buy themselves makes the cut. "If it has holes in it, if it's stained, if the zippers are broken, if the buttons are missing, if the collars are yellow, people are not going to buy it, and it's not saleable," Fred told me. I smiled and nodded, internally cringing at all the unsaleable things I've "donated" in the past.

On average, only about 40 percent of what we donate to the Salvation Army is even in saleable condition and makes it onto the sales floor. And of that 40 percent, only about 20 percent is actually sold. So to simplify, for every 100 pieces donated, only 8 will be sold on the Salvation Army sales floor; which means that $76,000 in potential sales is actually more like $6,000. (We'll get to what happens to the 92 other pieces momentarily, don't you worry.)

The garments that do pass the saleability test get a colored tag to help employees and knowing customers track how long the items have been in circulation. Each color gets rotated out every five weeks to ensure there's space for the new arrivals. On Wednesdays, which is Family Day, older colors get a 50 percent off promo. At week five, items of that color are half-off all week. In theory, then, you might score a pair of jeans, originally priced at $99.99, donated to the thrift shop where they're sold for $6.99, and further discounted to $3.49—less than a cup of coffee, or even an H&M T-shirt.

The tagging system might be the closest thing to a traditional fashion "season" in the donation market. But, unlike brands that have seasonal calendars, the Salvation Army will still sell a Hawaiian shirt or cargo shorts in January, making even off-season donations still viable for potential re-sale. Why? Major Fred told me that many of the people who are shopping at the store are immigrants who send the things they purchase back home. Moreover, since the store was mere blocks away from Manhattan's cruise terminal, many cruise employees came to shop there, too. (Cruise employees are often recruited from very low-wage English-speaking countries.)

As any thrift shopper knows, you can find surprising treasures in the Salvation Army racks. Major Fred told me they recently received a big donation of Chanel, which was treated like the fashion royalty it is. But the Chanel is getting choked out by polyester Zara tops.

The Salvation Army keeps prices low in order to move product, which is becoming harder and harder these days. It's become harder to sell used clothing in the United States for a few reasons. First, the clothes themselves. Fast fashion has a real expiration date, so buying an already-worn cheap top makes little sense, even if it's only for a few bucks. The most successful Salvation Army in the country is adjacent to a Walmart, suggesting they serve a similar low-income demographic. Most would probably choose a new $12 pair over a used $6.99 pair of jeans. The competition of new fast fashion is also just too big a force to be reckoned with. Major Fred scoffed at the idea of the Salvation Army going digital, à la ThredUp, but online shopping is where it's at.

But it's more than just the quality of clothes coming in that's different from when Major Fred started thirty-four years ago. Back then, people had two choices for where to donate clothes: the Salvation Army or Good-will. Over time, it's become clear that money can be made from our excess, so competition from for-profit companies has entered the scene—essentially, companies that play middleman and bring our discarded clothes right to exporters, or to resell to consumers. There's also more competition among charitable causes, too. Medical organizations—the heart association, the eye association, the kidney foundation, etc.—collect clothes to sell directly to sorters, to in turn fund their organizations. "A former boss got to the point where he would call them the 'body part boxes,'" Fred explained. I grimaced.

A baling machine at the Salvation Army in New York City.
Rejects come down the chute from the sorting floor down here where they
are bailed and shipped out.

The next stop for what's not sold is grading. The approximately 92 percent of clothing that the Salvation Army is unable to sell is baled up on-site and sold to graders, where it is subject to further scrutiny to determine its next destination. According to the Secondary Materials and

Recycled Textiles Association, 45 percent of these unsold donations is "re-used as apparel." Why is that in quotes? We'll get to that in the next chapter. Thirty percent becomes industrial and commercial rags. Even if clothes aren't good enough to be reworn, they can have second lives as rags if they're made of high-enough-quality material like absorbent cotton. Twenty percent gets reprocessed into its basic fiber content for things like furniture stuffing, insulation, and building materials, and the last 5 percent is trashed domestically. Garments that are wet, have mold, or are contaminated with substances are thrown away.

Grading used to be done domestically, but as has been the case for nearly every step along our clothing's journey, this process is also being outsourced with increasing frequency. Up to one third of all the clothes from Canada and the United States sent for grading lands at one location outside of the United States. Used Clothing Exports in Mississauga, Ontario, is a mecca of sorting (women's versus men's for example), grading (A/B quality or rags), pricing, and exporting, which amounts to 60 million to 70 million pounds of used clothing per year. But one goal of outsourcing, unsurprisingly, is to lower labor costs. The race to the bottom continues even at the end of clothing's life. Instead of paying American minimum wages for people to sort and grade garments post–Salvation Army, ungraded clothes get sent to places like Panipat, in northern India. Panipat has the highest concentration of clothing recyclers in the world. Another alternative is Pakistan, where labor costs several hundred dollars a month less than in the United States.

———

So now our little jeans, which we've probably forgotten about entirely by now and already replaced with a new pair, have made their way outside of the United States (again) for yet another trip around the world.

Historically, discarded clothing would only be good for a second life if reused as rags. New York, in addition to housing the original marketplaces of fashion production and sales, used to be an epicenter for rag dealers;

Brooklyn in particular was full of them, mostly Jewish immigrants (the Yiddish word *schmatta* means "rag"). Today the rag-making supply chain has gone international: a garment is made in South Asia, sent to and bought into the United States, donated to a thrift store, exported to India for grading, then sent on to somewhere like Star Wipers in Newark, Ohio, where some of the billions of rags used around the world every year are made.

If the garment is wool, which has little appeal to traders in warm equatorial Africa or Asia, it might get a special kind of reuse that's closer to what you'd think of as "recycling." Wool's short, coarse fibers can't be made into paper or furniture stuffing (or housing insulation, for that matter), and won't soak up grease or water like a cotton rag; so the wool gets torn apart and mashed back together into a "shoddy blanket." You'd recognize them from a medical drama, or coverage of a natural disaster, as most are used for emergency relief services like the Red Cross. Today, though, even shoddy isn't a viable recycling option. Because there are so many synthetics in circulation, wool isn't the only material that can be used for shoddy blankets; prices for secondhand wool have plummeted in the last fifteen years or so, from around 50 cents/kg to 15 cents/kg.

Despite these fairly standard and long-standing practices around donated clothing, other members of the industry—on the sales and marketing side of the consumption machine—are doing their best to make sure you don't know about it. Clothing brands, and their corporate social responsibility wizards, conceal the truth of what happens when clothes are donated. "Take back" programs at stores like Madewell and H&M encourage us shoppers to bring back our unwanted garments for the sake of saving them from landfills, or even explicitly explaining that they may be made into things like housing insulation—true, broadly. Here's the slant: When you do so, you receive a coupon toward a future purchase in return, in addition to "infinite good vibes" for donating, as Madewell tells its shoppers. Even Goodwill has their version: "we work with local agencies to recycle and re-purpose everything from electronics to books, and textiles to plastic toys, keeping tens of millions of pounds out of the landfills each year."

But when we zoom out and put this all in context, most of these take-back programs, just like donation centers themselves, are just another way clothes can make their way to a landfill. As stated on the website of I:Collect, a worldwide collection organization that partners with H&M, Levi's, Adidas, & Other Stories, Columbia, and others, their "innovative, cost-effective in-store collection concept that engages consumers, offers a reward incentive, *drives store traffic and sales*" [emphasis mine]. Madewell is not going out of their way to turn jeans into insulation, it's just the way it is, so they're exploiting that fact to entice you to buy new stuff. And anyway, those jeans you turned in to buy new ones were probably perfectly wearable in their original form. In 2018, H&M collected 22,761 tons of unwanted clothes with the messaging that this was a special initiative on their part. Its website presents three possible futures for your donated jeans—they might be reworn as clothing, reused as "remake collections or cleaning cloths," or recycled as textiles for the purposes listed above. And you'll get 15 percent off your next purchase for a whole bag of 'em. What's left unsaid is that in reality, 50 to 60 percent of those donations all met the same fate as any other donated garment—exportation. And don't forget that as recently as 2017, H&M was burning its own unsold inventory.

None of this, of course, is under the purview of the Salvation Army or any similar thrift shop. They, too, wear the out-of-sight, out-of-mind-colored glasses even when it comes to their own business. Once they get paid for their unsaleables, whether those clothes find their way to a closet in Africa, become a rag, go into home insulation, or just end up directly in a landfill has no impact on the "army" of sorters and sellers.

When I asked Fred about his own relationship with clothes, he looked down at his uniform of dress shirt and slacks. "I don't buy anything. I don't like to shop," he said with a chuckle. That told me enough about how his work had shaped his opinion about what we wear, and what we don't.

Eight

Paved with Good Intentions: The End of the Road for Clothing in Ghana

C ause I am beautiful, I got beautiful things." Abena's melodious voice mingled with the tinny percussion booming from a nearby speaker, a restless beat that stayed in my head for weeks. A retailer at Kantamanto, one of the largest secondhand markets in all of West Africa, Abena gestured dramatically around her neatly organized but utterly packed stall. "Office shirts, casual shirts, dresses, long tops, jumpsuits—I have *everything,*" she explained. I didn't doubt her for a second. Hundreds of ladies' garments were hanging from the walls of the stall, piled on the floor, and tied into a wheel so the clothes bloomed out like a many-textiled flower. It was very different from the neatly labeled racks of a Western department store. But here in Kantamanto, devoted to selling the donated and discarded clothes of the Western world, anything goes.

Kantamanto sprawls over six acres in the main business district of Ghana's capital city, Accra. It was overwhelming to say the least. Thankfully, I had a trusty guide. Liz Ricketts ("Lizzy" to her Ghanaian friends), founder of the NGO The OR Foundation with her partner, J. Branson Skinner, has spent the last three years navigating the five thousand stalls of the market, getting to know some of the thirty thousand Ghanaians who find work there as retailers, porters, and more, and surveying the enormous spectrum of clothing that passes through the market every week. Liz had given me a few pointers during our half-hour walk from our Airbnb to Kantamanto: Sellers would be very aggressive, but since it was Thursday—market day for retailers—it might not be so crowded. Wednesdays and Saturdays were when customers descended—ten thousand per day. Having navigated my way through Bangladesh, Sri Lanka, and China in past months, let alone the sometimes vicious retail centers of New York and Middle America, I felt ready—excited, even—to see what Ghana had to offer its shoppers. How intense could it be?

About fifteen minutes into our walk, I started to realize I wasn't in Kansas (or Minnesota, or Dhaka, or Manhattan) anymore. Along the side of the road were street vendors hawking goods of all varieties—clothing, lingerie, and shoes, but also yams, eggs, tomatoes, onions, and ginger. Distracted as I was by the variety of colors and smells, and voices crying out prices—"three cedis, three cedis, three cedis" harmonizing with the increasingly loud drum music streaming from anywhere (or everywhere)—I had to stay focused on where I was walking to dodge everyone else. All around were women carrying enormous loads on their heads—metal baskets and bales of secondhand clothing to bring to market, or bushels of produce—or babies on their backs. Frequently, women were laden with both. Some of these women are *kayayei*, or "girl porters."

I was flooded with memories of my first trip to the Javits Center for Zady research all those years ago. At the time, I'd never seen so much clothing in one space—not even in my crowded closet or the flagship H&M I'd frequented during my law-firm days. But Javits had nothing on Kantamanto, or really any shopping Mecca in New York City including the multistory flagship locations of fast fashion brands or the packed racks of the Salvation Army. An estimated 2,425 tons, or 15 million items, pass through the market per week. Among its ten thousand daily shoppers are people from all over the country as well as people from nearby West African nations such as Nigeria, where secondhand imports are illegal.

I was pulled out of my reverie by a sharp cry I knew was meant for me: "Hey, *obruni*!"—hey, foreigner (or just white person, depending on whom you ask). Very few of those shoppers look like Liz and me; during my five trips to Kantamanto, I think I saw two other white people, total—one of whom asked where to buy African art, the equivalent of going into Zara and asking where their spaghetti aisle is.

Following closely behind Liz, I started to realize that Kantamanto, while initially unfamiliar to me, was really no different from another mall or department store, just at many times the scale and partially open-air for circulation and light. The market was divided into sections—men's, women's, kids', and subsections therein of denim, suits, blouses, dresses, and so on. There was also a food court, a salon section, and aisles of tailors and seamstresses.

If I went straight down the street, I discovered, past the woman selling African print baby clothes and then past the sneakers, and then looked out for a gray building, I could hang a right and walk down a small hill with street hawkers selling mostly men's clothes and soon I'd be back on the main market street. "When you notice the smell of chickens," Liz's voice echoed, "look for the man selling winter coats, then take a right there under the covered market. You will know you are on the right path if you are in the men's area and you see the stall selling exclusively white shirts."

For many Ghanaians, getting around Kantamanto is second nature. The secondhand market is a go-to, baked into Ghanaian culture. Sold at rock-bottom prices—local YouTube vloggers boast shopping hauls that show off the jeans they got for 12 cedis ($2), or the five shirts they got for 7 cedis—and in endless varieties of patterns, colors, cuts, and textiles, these select garments keep Ghanaians on the cutting edge of contemporary Western fashion, or whatever style suits their particular taste. There is no shame here in secondhand clothing, unlike in the West. Thirty million Ghanaians—90 percent of the population—from every income bracket buy secondhand clothes. Kantamanto is a way of life, and an exceptionally stylish one, as I would discover.

Still, trawling the busy marketplace isn't everyone's cup of tea. Some upper-class Ghanaians with money to spend might hire personal shoppers to fetch specific styles from Kantamanto; others prefer shopping in smaller secondhand boutiques, where shop owners have already combed through Kantamanto's stalls to curate a selection of wares. The most up-to-date sellers have migrated to Instagram, where they post about "upfitted" and "upstyled" goods from Kantamanto. They also up-price them, of course, and deal with customers via direct messages. The less well-to-do might shop from "mobile sellers," who take up a selection of clothing and carry them to the farther outskirts of Accra and beyond. Some sellers might even get a gig with a repeat client as a personal shopper. Only the ultrarich don't shop at Kantamanto, including, ironically, some of the importers of secondhand clothing. They buy their nonsecondhand clothes in Europe. I met a Ghanaian photographer and law student—his mother was a judge—who had never been to Kantamanto, only to the mall.

Shop sellers in Kantamanto tend not to be picky about who buys from their stock, whether it's the boutique owners buying in bulk, Instagrammers, regular people, or hawkers. Inventory is just not the problem. Every week, one hundred forty-foot-long containers, each carrying about four hundred bales of secondhand clothes and weighing 55 kilograms (121 pounds), arrive at the port in Tema, about an hour's drive from Accra and twenty minutes from the Kpone landfill. These are the same bales that were graded in Canada and India. Each bale is sorted and labeled by gender and product category: for example, women's tops or men's suits. Bound in the thick plastic, they're also labeled with information about the wholesaler that sent them. Bales originating from Europe and America are preferred, as they are considered to be of relatively higher quality—a survey found that British women wear their clothes only seven times—compared to secondhand or even new garments coming from China. But just as many Westerners are unaware that their donated clothes are sold to the African importers in bulk by graders and then resold in markets like Kantamanto, many Ghanaians have no idea that the clothing coming to them was given away for free.

As I walked through Kantamanto that first Thursday on the morning of a delivery, the bales were formidable and impossible not to see—stacked two or three or four high on hand trucks on the outskirts of the market, bobbing around the market masterfully balanced. The bales seemed to navigate through the market like ghosts.

Bales of secondhand clothes on their way to Kantamanto.

But what I saw was only part of the story. In the wee hours of the morning, crews of young men crowd outside of Kantamanto, transferring the bales from the trucks that brought them from Tema into importers' storage facilities. A *kayayei* brings each bale to the sellers inside the market itself. Each woman might take up to ten trips back and forth to the market, earning between 1 and 3 cedis per trip. Since most action at the market—any day of the week—starts around six a.m., and the bales need to be delivered, unpacked, and sorted by the time the early-bird customers arrive, this generally happens at the crack of dawn.

Waiting eagerly for their deliveries on Thursday morning, the sellers usher in the bales, usually about five per week, and pause to pray that a good lot landed with them that day. Sellers have no clue what could be inside their bale of "ladies' tops" they'd preordered, sight unseen. Since the contents are occluded by the opaque plastic and the bale cannot be opened before purchase, sellers take on enormous risk. A bad bale can bankrupt a seller. A good one can pay for a child's or sibling's school fees for the week. Sellers waste no time in slicing through the thick zip ties that weave a neat checker pattern around the bales. Freed from their opaque plastic cages, the clothes spring to life once more on a shore thousands of miles away from their last use.

Bales look equal on the outside, but inside is a whole different story. The seller's job is to sort—yes, again—the contents of the bale into three categories: first, second, and third selection, a ranking system based on quality and, as I learned, a lot of subjective fashion savvy. I was schooled on the art of sorting by David Adams, a stall owner who specialized in men's suits. David started selling at Kantamanto in 2013, when he was a third-year at university studying mathematics. The market had just suffered a devastating fire. His family begged him to come back from school to help support them as they rebuilt their stall. Temporarily setting aside his own education, he worked to pay for his younger sisters' schooling; one was at the University of Ghana, and the other at the international high school. David started by hawking second- and third-selection items from bales he purchased. After just a week and a half, he was buying bales of suit jackets for 150 cedis ($26) a pop, selling to his father's existing clientele at the market. In two months' time, he had enough money for his sisters' tuition and was able to go back to finish his own degree.

But all throughout his fourth term, he kept getting calls from old clients who said they needed him because "he could relate to them"—he was a real-life influencer. The suits themselves seemed to call him back to Kantamanto. During David's year-long national service period after his graduation, his meager stipend trickled in in unreliable increments, making Kantamanto comparatively attractive. The suit bales were where the money was, he knew. This was sometimes literally the case. Old suits used to come in with dollars forgotten in pockets; "now nothing comes, only a tissue pack," he said with a chuckle. He didn't need a math degree to understand that if he bought a bale of suits for 600 cedis each, and sold just two jackets at 150 cedis each, he'd cover his costs and be able to use "the rest to buy a Microsoft phone."

David's day starts at six a.m. He spends an hour and a half cutting new bales and sorting sizes and grades. The first selection is the cream of the proverbial crop. These are the clothes that look like new, or are new with tags. Even Goodwill tags count; they dangled from many items I saw in the market. First-selection clothes should be fashionable or have a well-known label sewn in them. He held up a slim-cut suit as an example; David also looks for pockets that are still sewn shut, indicating that it was never, or just occasionally, worn.

About 10 percent of his bales these days are first selection. Back in 2008, it was more like 50 percent. Then, the pockets also used to be stuffed with cash. Now, the majority of the bale, 60 to 70 percent, is second selection—garments that are fairly worn but may still be desirable to someone. Clothes that are faded, worn out, or older designs—a pilling three-button suit, for instance, or a stained T-shirt with a stretched-out collar—are third selection, which David sometimes manages to sell but more often end up directly in the discard pile, called *asaei* (under) or *bola* (trash). David pointed to a woman's dress as an example. He looked at the label and then to the zipper. "Look at this," he said, with mild disgust. "It looks like something my grandmother would have worn." It was true: The label and zipper looked vintage, but not in a good way.

Increasingly, the "under" also includes deadstock from clothing stores that is deliberately slashed to make it unsellable. Third-selection pieces also include those that don't suit the needs of Ghanaians. They want breathable fabrics in this hot and humid climate, so garments made of fabrics like silk, which wrinkles easily and is not great at hiding sweat, are unsellable. Larger sizes, which are showing up in more bales from obesity-plagued Western countries, are also unsellable. According to Major Fred from the Salvation Army, Japanese garments are more popular now for the secondhand market as a result.

Of the perhaps six pieces that are third selection, David might be able to sell four. Suits are unique in this way—there are always people who buy them for the deceased to be buried in, in which case style doesn't matter so much. Fitting: the Ghanaian words for these secondhand clothes, *obroni wawu*, translate to "the white man is dead." It was once so inconceivable that one would just give away a precious commodity that it was assumed the person had actually died. That's how precious clothing used to be and still is for some people.

Not all sellers have the same proportion of selection. Suiting is the best category in terms of sales and margin. Remember that stalls specialize in types of garments and a certain gender; Liz's NGO recorded a denim seller sorting through one bale that resulted in eighty-two third selections, thirteen second, and just eight first selections. On the other hand, denim is unique among secondhand clothing because second- and third-selection pieces can be renewed. Thanks to nearby tailors and dyers, jeans can be refreshed for a more desirable look; Ghanaians prefer dark and more structured jeans, since they're more formal-looking, in contrast to Americans' Levi-cowboy slouch style.

No one on the Ghanaian side has control over what the bales contain. Importers, who order bales from graders, typically have a lot of capital in order to invest in the scale of containers. Many have been in the business since the 1960s, and understandably so: Liz shared that successful importers stand to make more than $100,000 (yes, American dollars, not cedis) per week. Still, they have no say over what comes to them in the weekly delivery and no negotiation power if the quality is low. But they don't actually bear that risk. That gets passed down to the shop seller.

The average bale costs a shop seller about 937 cedis ($165) and brings in an average gross profit (sales minus the cost of the bale) of 478 cedis ($84). That money is then used on business expenses, like hiring porters, paying rent on the stall, cleaning, and cell phones, which might add up to 100 cedis per week ($18). Shop sellers depend on first selection to make up to half of that income. Because first selection is in high demand, it might sell out before lunch on a busy Wednesday; that leaves Thursday and Friday to sell the remaining clothes and make up the rest of their earnings. Whether or not they hit that mark all depends on what the mystery bale contains—and on who's doing the selling. According to Skinner's research for his master's thesis, men, who mostly sell men's clothing, on average make a profit at the market, whereas women sellers, who mostly sell women's clothing, on average actually lose money. This is due to a variety of factors. The women's area is older than the men's area, so it's not as attractive a shopping experience (it's extremely crowded), and more women's clothes are imported, which may translate to more clothes of lesser quality that are less likely to sell. I would add a third reason: that women perhaps have less spending power than men do.

David is in the right business—suiting—and does an excellent job in sales. He makes a 50 percent gross margin on the average bale, which explains why he stays in the business despite its not being seen as particularly prestigious in Ghana. This is no quaint yard sale, but a booming economic hub. But of course not everyone at Kantamanto reaps equal benefits—not just less lucky female sellers, but also the shop sellers' subcontractors. Part of David's expenses go to hiring workers; he'll pay three men 30 cedis a day for unloading and other tasks, and 50 cedis on Saturday since no one works on Sunday, and they appreciate the extra cash when they go out at night. The workers make enough to live, not to save. But there's a clear desire from everyone involved—the importers, the shop owners, and the workers—for the government to step up its investments in Kantamanto, given how central it is to people's livelihoods and the opportunity involved. David's future aspirations range from continuing his education, to opening a factory for suits, to growing mushrooms and snails (he likes biology). I suggested he'd make a superb political leader. But for now, he says, "I prefer to do this business."

"SLAVES OF THE SYSTEM": THE *KAYAYEI* AS LIVING HISTORY

One group of Kantamanto's most essential workers has been shouldering the functionality of the market in the most extreme and literal of ways: The *kayayei* are the invisible heroines, and victims, of the end-of-life stage of clothing.

The fleet of women buzzing through Kantamanto on the first day of my visit, with 55-kilogram (121-pound) bales on their heads and babies on their backs, left an impression on me, but to everyone else, they were just an accepted, normal part of the market.

The *kayayei*'s main job is to carry the bales of clothes from the exporters to the retailers' stalls in the market, but they can porter just about anything. Wealthier shoppers may hire a girl to carry their day's loot home, and others spend their days toting vegetables to the people who live more remotely at the farther edges of towns and villages. Tomato sellers have one of the longest commutes among all the porters. Their loads average 130 to 200 pounds, often more than the women's own body weight. The few male porters take on the 200-plus-pound loads (anything bigger gets carried by truck). The porters travel great distances on market day. They sometimes travel a whole kilometer just within Kantamanto as they weave through the very narrow, crowded aisles, which are navigable only by foot, rather than by car or even hand truck.

Watching them load their bales is like watching a mix between ballet and deadlifting: After securing a small ring of fabric to their head for a cushion, they squat down so another woman can hoist up the bale or a metal tub filled with goods. The porter then stands straight upright, and is off. One *kayayei* named Asana told me that for all this she earns 10 cedis per week. At max, on a busy week, she would earn 50 cedis per week, less than $9 US. In addition to this compensation, girls are paid (with interest) in neck and back pain, including some cases of women breaking their necks.

Girls like Asana take many roads on their way to Kantamanto. Most come from the north of the country, choosing work as a *kayayei* as an alternative to being sold off in marriage, or to earn money so they can find a man to help support them if their husband or father has died. Climate change is also wreaking havoc on Ghanaian agriculture, so many girls are also sent away when their family farms go under from flood or drought. Just as in Bangladesh, those most affected by climate change become the most vulnerable workers in an industry creating climate change.

From these places of fermenting distress, the *kayayei* land in a part of town steeped in suffering. Old Fadama, which is about half a mile away from Kantamanto, is not only the oldest slum in the country but is also directly adjacent to the world's largest electronic waste dump, Agbogbloshie. As in the United States, the marginalized members of society here are also often forced to live near toxic waste. The girls live ten or twelve to a room, arranging themselves foot-to-foot and head-to-head to sleep in the suffocatingly hot and small rooms. They get just one key for the house, meaning that everyone else must wait for the girl who has it to come home. There is no margin for error, no amnesty. If the rent isn't paid immediately on time, everyone is kicked out. The vast and intense pollution and living conditions are so bleak that the area has been given the nickname "Sodom and Gomorrah," which were the precise words that entered my head upon visiting.

Their days start around 4:30 or 5:00 a.m.: They need to get to the importers' area on the outskirts of the market and then arrive at the market by 7:00 a.m., when it opens and unbaling begins—a schedule not dissimilar from Amazonians'. Although it is tireless, thankless work, the *kayayei* are so indispensable to the market's functioning that they know they are guaranteed to be rehired every week. Some might argue that's better than nothing, but "they are the slaves of the system," as Liz says.

During COVID-19, the *kayayei* were among the hardest hit, given their crowded living conditions, lack of sanitation, and distance from formal medical care. On top of the threat to life itself, market closures meant that what little income they typically earned disappeared.

During one visit to Kantamanto, I found myself at a busy juncture clogged with trucks and people. One woman had a wide metal tin on her head, with the biggest yams I've ever seen arranged in a neat circle for easy picking. Another stood with a pile of T-shirts folded and tied to her head, a piece of string snug under her chin, and hangers with magenta polka-dot and leopard-print blouses hanging off her own chest. Down the road I could see more bales bobbing, their human carriers invisible beneath them.

What were they thinking about? I wondered. Were they like Danu and the other Sri Lankan garment workers I met, who had also moved to Colombo from the north for the menial work society gave them little choice but to perform, and did nothing to protect them from? How many more women are there around the world, at this very moment, performing some work in the service of our fashion, paid or unpaid, respected or taken advantage of, legal or illegal? In this place where clothes go to their graves, can we find ways to give people a chance for life?

TALKING TRASH: WHY WE NEED TO REEXAMINE THE SECONDHAND MARKET

Kantamanto has a visceral energy. Incessant drums beat out a rhythm for shoppers as they move from stall to stall and sellers' shrill voices meld into a surprisingly harmonious chorus as they hawk their wares and their prices. And shoppers seek treasure like looking for a golden needle in a haystack. It would be easy at first glance to think it's a great big party. And yet, the whole reason Kantamanto exists is because of the gross overproduction and undervaluing of garments we've been exploring throughout this book. This is becoming a global problem, but the United States is leading the charge. Every year, the United States exports more than a billion pounds of used clothing. That figure, of course, is only a portion of the global trade. Data from 2016 shows that of all the world's secondhand clothing, 40 percent comes from just three countries: the United States (15 percent), the United Kingdom (13 percent), and Germany (11 percent). Together the EU and US represent 65 percent of the value of global used clothing exports. As countries become wealthier, they export more clothing. In China, the player to watch in every race, used clothing exports doubled from 2006 to 2016, an interval in which its middle and upper classes grew significantly. As we've seen, clothes make their way to Ghana from a variety of places: the

Salvation Army, give-back programs at stores, as well as unsold, unworn deadstock from retailers.

COVID-19 interrupted this cycle on many levels, for good and bad. Remember how we talked about the decline in donated goods as the Salvation Army and other such centers were temporarily closed? Between the reduced supply, the closing of borders to block the spread of the virus, and social distancing efforts, markets like Kantamanto were shut down almost overnight. According to Steven Bethell, president of Bank & Vogue, a grading company in Ottawa, Canada, prices for clothing on its way to Africa fell by more than half in spring 2020. Its destinations refused to accept as much as it previously had. Kenya temporarily banned clothing exports entirely, and Mexico, which takes at least 30 percent of American thrift stores' unwanted donations because of its proximity to the United States, was closed off due to travel restrictions.

The breakdown of the system due to COVID-19 demands that we examine how and why it was so broken in the first place. Part of it has to do with our consumption habits, some of which we justify thanks to our friends on marketing and corporate social responsibility teams and in donation centers who craft appealing language around "take back" programs, as we saw in the previous chapter. Others argue that donated clothes create jobs in the developing world. People like David and Abena and the thirty thousand other retailers are only able to get by because of their jobs at Kantamanto, so why would we take that away from them?

Yes, reselling creates opportunities for work. But the kind of profits David brings in are not common. Remember that many female sellers lose money, often resorting to loans extended by the local importers. Now it seems less like a job and more like debt bondage. And the secondhand market blocks other opportunities that could be more stable and resilient for local economies. The secondhand market keeps local economies at the mercy of more developed countries. The Trump administration even suspended Rwanda's beneficial trading status through the African Growth and Opportunity Act (AGOA) when it banned imports of secondhand clothing, as it attempted to grow its own domestic industry. As a result, Rwanda's clothing exports to the United States are no longer duty free, a punishment for not happily accepting our trash.

Sammy Acquah from Akosombo Textiles Limited (ATL), a Ghanaian manufacturer of its famous traditional waxed cotton, explained that customers' more fickle style preferences these days, which are enabled by the variety at Kantamanto, are lowering his business's sales. He sees a bigger threat, though, from knockoffs. Since China joined the WTO in 2001, he says, he has noticed that some Chinese businesspeople will come to textile stores, like his, buy an original, 100 percent cotton textile designed and made in Ghana, then go home to re-create it with subtle deficiencies. The fake "Made in Ghana" tags are never placed quite right, the fabric feels cheaper and less soft, there are inauthentic design numbers on the tag, and, most important, they cost a quarter of what Sammy's cloth costs. As a result, China has captured about 50 percent of the textile market in Ghana.

Taken together, these factors make it clear how and why for the last several years, secondhand and cheap imported clothing has been steadily replacing Africa's once-vibrant textile industry as a domestic and exporting trade. In stark contrast to France during the time of the Sun King, the US and UK during industrialization, and China during the open-door policy, the developing world today has not been able to use the tool of protectionism to develop its own industries. Why? Neoliberalism. Beginning in 1989, the Washington Consensus was a neoliberal policy by the United States, the World Bank, and the International Monetary Fund to ostensibly open trade to developing countries by forcing them to remove trade barriers for imports. These policies essentially required developing nations to become importers of our goods—and trash, as it turns out—rather than allowing them to foster strong domestic industries that could produce goods for export. As a result, these economies became tethered to the superpowers of the world, making them especially vulnerable to changes in the dollar exchange rate—or a closure of international borders.

The future at the moment doesn't look very bright for the retailers in

Kantamanto. On average, only about 18 percent of bales today are first-selection garments, on which retailers need to make 50 to 90 percent of their money back. The more disposable fashion we give away, the more pressure there is for retailers to sell fewer garments for more. Not only are these not high enough quality to last into a second or third life, but they're not the kinds of styles or fabrics that will appeal to Ghanaians, who prefer more structured and formal-looking clothes—attire with seams and tailoring, not stretchy knits.

Trash to ... Trash: The Landfill You Didn't Know Existed

With a shrinking inventory of quality secondhand clothing and a growing inventory of disposable clothing, Ghanaians are forced into doing exactly what we're led to believe *won't* happen when we donate or give back our undesirables. They travel thousands of miles only to get thrown

Me at Kpone, the landfill where 77 tons of textile waste from Kantamanto gets dumped per week.

away in the trash, never even glimpsing their promised second life. What's more, the disposal systems in Ghana and much of the developing world are less developed than what we saw in New York, and throughout the United States, which results in more pollution and climate impact than if it had just been thrown out in the developed world, to say nothing of the environmental cost of transporting all that waste.

Trash is indeed as integral to the hubbub of Kantamanto as fashion is. Built into the unbaling process is an hour, around 5:30 to 6:30 a.m., when all of the unwanted third-selection and under gets swept up for the trash. Then retailers have to deal with what doesn't sell. According to Skinner's observations, the Accra Metropolitan Assembly (AMA) collects 77 tons of clothing waste six days a week—2.8 million items per week—from Kantamanto and disposes of it in Kpone, the main landfill. For every three garments sold at Kantamanto, two get trashed. Let me pause here a moment. More than half of our clothing that is sent to Accra goes to the landfill. We need so much more data than one person's graduate research, but this startling statistic upends the entire point of the whole resale and donation system.

A pile of under inside Kantamanto. Each intersection of the market accumulates its own mound of trash.

*The truckful of under—from the market trucked
back to Tema, only this time for landfill.*

This turnover has made Kantamanto the most consolidated point of
waste pickup in all of Accra, and possibly in the whole country. If the
average landfill in the United Sates contains about 5 percent clothing
and shoes, Kpone is four times that—a full 20 percent of the entire
landfill's planned capacity is taken up by Kantamanto's waste. Still, only
about 25 percent of Kantamanto's total waste is sent to a landfill. Another 15 percent is picked up by private, informal collectors who may
illegally dump it in waterways, bury it on beaches, burn it in open lots,
or simply leave it along the side of the road. This unregulated dumping
was behind the 2014 cholera outbreak, which killed 243 people. Kpone
opened in 2013 and exceeded its full capacity in under half the projected time frame—six years instead of the twenty-five-year projected
lifespan.

    In a sad way, this final leg of our clothing's journey makes complete

Trash along the side of the road surrounding Kantamanto.

sense. First, labor and production were sent overseas to countries where the absence of regulations made everything cheaper, and kept the realities of making our clothes invisible—and dirtier. Now the same thing is happening with our castoffs so that countries without reliable infrastructure drown in our garbage.

Excessive textile waste from halfway across the world is becoming an increasing burden on a city that's already in dire straits. Clothing is burned in people's backyards. It clogs the city's sewer system and fills up the city's dump at a truly unsustainable speed. It is terrible for citizens and, unsurprisingly, the planet, as I would learn on my visit to Kpone.

———

We rolled up to the landfill on a hazy Friday morning. While glamour was clearly not the theme of any of my travel to report this book, I was explicitly briefed to not wear eye makeup to Kpone; the chemicals in the landfill would make mascara congeal on my eyelashes. Adjacent to the entrance, at the edge of a shantytown (evidence yet again that the most disenfranchised suffer the most from waste), a few men were engrossed in a checkers match. Turns out the haze wasn't just weather, but smoke. The landfill was on fire.

I had in fact seen the smoke from many miles away, but assumed it was something else. The closer we got, though, the clearer the source became. About a quarter of the landfill was smoldering, creating clouds of black smoke that billowed as far as the eye could see. The sky above the rest of the landfill remained incongruously clear and blue.

The scene on our arrival to Kpone.

Stepping out of our car on the burning side was like walking onto the set of *The Hunger Games*. I looked around and remembered the places I'd visited while researching this book: the fields in Texas, where pesticides floated through the breeze along with sprays of cotton fluff; the chemical-slick river in China I almost fell into, and the jeans factory floor where I stood in a puddle of effluence; the side streets of Dhaka where bolts of forgotten fabric mingled alongside felled trees and tires and discarded shoes; the slums of Sri Lanka, where a dozen young girls were herded together to spend the end of their childhood fervently working on an assembly line while they sewed lingerie. Each time, I thought to myself: *Now, this is the dirtiest place I've been.* Here in Accra, I had already spent most of my time in my room just trying to keep clean: As soon as I'd swept the floors and washed my feet, I would take a few steps around and find my soles covered again in a brown film. Liz explained that it was probably because our Airbnb was near Agbogbloshie, the electronic-waste dump near where most of the *kayayei* live. And now, standing amid the smoke, the roaring wind, and the increasing heat, I was experiencing the truly dirty.

I looked down at the dense clay earth to try to regroup. But when my eyes landed on the outline of a pair of jeans and a child's embroidered dress embedded into the dirt like fossils, my mind transported me to all

the places those clothes might have been—Was the cotton grown in Texas, or India? Were they finished in Dhaka, or Shaoxing? Were they sold in New York, or Minnesota, or London, or did they pass through Kantamanto?—only to come here to die this shameful death, the people whose hands made and sold them completely forgotten.

We had arranged to meet up with Percy, the manager of the landfill, though he was naturally sidelined by the fire that was smoldering in front of us. "This is very embarrassing for me," he said as we asked if we still had permission to explore, "but it's the reality." He decided to grant us access to the landfill anyway in the spirit of authenticity. But not everyone thought it was a good idea to let a bunch of outsiders tramp through burning garbage unsupervised. Once we ascended the trash mountain, a landfill worker came running after us, shouting, "You can't be here!" Once he cleared up with Percy that we were given permission to explore, he was very nice. With courteous anxiety in his voice, he shared that it was the worst fire they'd ever had—fires had clearly happened before, but they didn't seem regular. They thought the spark might have come from the company next door. The next day, media reports suggested that the fire might actually have come from embers inadvertently brought in with other trash.

Climbing through the landfill was equal parts treasure hunt, horror show, and *Fear Factor* as the smoke descended. Although I wore a face mask and deliberately protective long sleeves and pants, the murky morass of food scraps, plastic bags, and of course clothes was alarming. I kept my focus by playing a game of I Spy. Plastic bags were the most easily identifiable, their bright neon blues and greens dotting the side of the otherwise greige topography like confetti. My black knit Nikes found all sorts of mates—swoosh-bearing sneakers and slides, along with Asics and a pair of red Velcro Puma high-tops. Other clothing was harder to identify, having been mostly crushed into the dirt like those somber jeans or packed in the trash bags, so that only the brightest neon clothes stood out.

Each time I recognized something, I paused. A sorbet-plaid button-down with an H&M L.O.G.G. (Level of Graded Goods, a laid-back beach/outdoors line for men) label. Another tag proclaiming "Made in / Fabrique au Canada" stuck out of the heap. A pristine-looking white faux Versace dust bag lay in stark contrast to the rust-colored soil. A canvas tennis shoe that had been so mangled it was barely recognizable.

I had fully expected to be suffocated by the stench of garbage. And yet surprisingly, the aroma of Kpone was tolerable. Don't get me wrong, it was still a mountain of trash, but the fire somehow mellowed it all out to an unusually pleasant, bonfire-esque aroma—you might call it burning, with top notes of decay. The odd coconut or banana peel or Styrofoam food container peeked out beside a flip-flop or plastic bag. But there was much less exposed food waste and fewer maggots than I had expected because most of the trash was contained in bags.

Every so often we'd come along other trash-climbers—people who were unofficially working at the top and down the side of the landfill. They had a much more focused goal. When the trucks dump their contents, they sift through, pick out the plastic bottles, collect them in enormous bags, and roll them down the hill. They could make up to $20 a day with this work, though it was unclear who paid them and for what reason. (I could not find evidence of any recycling facilities that might want to buy the bottles.)

Once I got to the top of the hill, I stopped to survey my surroundings. There was something ugly-beautiful (I feel like there must be a German word for that) about the plumes of black-gray smoke extending out into the horizon. We couldn't stay too long at the garbage summit as the flames approached. Because the landfill had already exceeded its capacity, in less than half the time projected, as I mentioned, certain critical safety measures had been abandoned. For example, the landfill had once been separated into four quadrants. No more. Which is why what could have been an isolated incident in one quadrant was now beginning to burn the entire landfill. After spending a few good minutes at the top, letting the approaching smoke clouds wash over me as I surveyed the graveyard of clothes under me, I decided it was time to descend. The heat was intensifying at my back and I recalled my now young child waiting for me. *Well, this would just be a really stupid way to die,* I thought.

≡

Liz had pointed out to me that the do-gooder narrative around secondhand clothing in Africa skips over a key truth that I saw with my own eyes. For every campaign about donating or giving back clothes, there should be five more illustrating how much work goes into actually giving these clothes a new life *and* how much of it actually just gets dumped. If the ratio of first to third selection continues to diminish, Kantamanto will lose its vibrancy and socioeconomic value. However, if Ghana and other developing countries do continue to keep the secondhand trade open, there is a straightforward way to improve the system: sending higher-quality, more durable clothes. If the secondhand garments entering Ghana were of actual high quality, Ghanaians would have something they, too, could wear and rewear. Fewer garments in the trash—whether we do the tossing or West Africans do—would also shrink the overall environmental impact of our clothing. When trash burns in an open environment, like what happened that day in Kpone, or what happens every day in unofficial waste fires, its solids are converted to gases and particulate pollution. You know LA's characteristic haze? That's particulate pollution. The gases include climate-change-causing greenhouse gases like CO_2 and methane, as well as toxic gases like formaldehyde and acetaldehyde, and carcinogens like butadiene and benzene. Since synthetic clothing contains more carbon to begin with, when it is burned, it releases more CO_2. Just as the washing process causes the release of plastic microfibers, when synthetic clothing is burned the plastic microfibers are also released into the atmosphere and end up wherever the wind takes them.

All this is to say, it makes absolutely zero sense to put so many resources into producing a garment, sending it halfway around the world to be sold, wearing it only a few times, and then having it sent halfway around the world again for it to just end up spewing all of those resources up in the atmosphere and into people's lungs, soil, and waterways.

But whether they're burning or not, landfilled textiles are always a source of greenhouse gas emissions. Rewearing trumps both recycling and disposal of clothing: Wearing a garment twice as long would lower greenhouse gas emissions from clothing by 44 percent.

When I got back to my room that night, mentally and physically drained and longing for a bit of escape, I pulled out my phone and started scrolling through the news. (I'd never guess the news would be a welcome respite to anything.) Kpone was not the only place on fire that day, I learned. A slum in Bangladesh, inhabited by mostly garment workers, had also erupted in flames.

Rima's tin-walled dwelling in Korail was a good 6.5 kilometers east of Chalantika, where the fire broke out, but I could just as easily have been reading about her and her neighboring garment workers. Ten thousand people would be rendered homeless by the blaze, which consumed the little they could call their own. Fires in Siberia had been breaking news just weeks before I arrived in Ghana, and shortly after I got home the world watched aghast as more than 2 million acres of the Amazon rainforest burned to the ground. By January 2020, as I started writing this chapter, unprecedented bushfires claimed 46 million acres of land in Australia, and the lives of more than one billion animals, putting some species on the brink of extinction. The world was on fire.

THE ALTERNATE ENDING

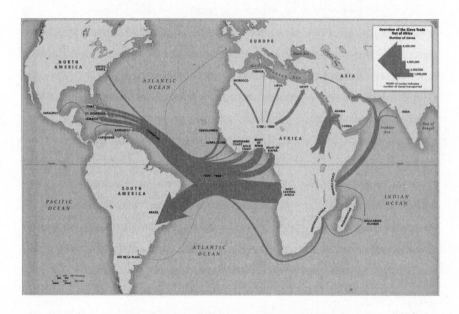

The trade lines of used clothes are surprisingly similar to former trade lines of people who were enslaved/cotton/sugar.

I signed up for a bike tour of Accra on the second-to-last day of my trip. It was a Sunday, the only day these tours are held. Because almost everyone goes to church, the roads that are normally thrumming with music and voices and car horns become quiet enough to make biking relatively safe. After about thirty minutes, we arrived at one of the tour's main attractions: the famed Osu Castle. Built in the mid-1600s by the then-kingdom of Denmark-Norway, the castle played a pivotal role in Ghana's relationships with the Western world, which centered on clothing. During the heyday of imperialist land-grabbing and trade, it passed through many European hands and served as a fort and site for bartering. Thanks to local chiefs who served as middlemen, the Danes were able to establish a trade monopoly along the entire west coast of Africa. From Europe came guns, ammunition, liquor, cloth, iron tools, brass objects, and glass beads, and out of Africa came gold, ivory, and people who were enslaved. This human property was at the heart of the entire trade system after 1697, delivering people who were enslaved mainly to the Danish West Indies or selling them to other European nations. Between 1660 and 1806, between 100,000 and 115,000 people who were enslaved were transported out of the castle.

After the transatlantic slave trade was abolished in 1808, and later after the British took control, the fort fell into disrepair and out of function. It served as the Ghanaian White House for some time, but now it mainly holds a clinic, shops, a café, a post office, and gardens open to the public.

Our tour guide, Samuel, showed us some of the places where Africans would be pressed into enslavement. Just twenty miles away from here, where the people whose labor in cotton fields fueled the Industrial Revolution and our modern economy were first chained together, our *obruni w'awu*, our dead white man's clothes, landed. Clothing connected America and Ghana during one of our darkest times in history. And clothing still threads us together today. How it connects us in the future, for good or bad, is up to us.

———

A neat, color-coordinated display of clothing stood against the back wall of Stefania Manfreda's pop-up concept shop in the trendy neighborhood of Osu in the heart of Accra. The fabrics were as breezy as the air blowing from the air-conditioning unit along the side wall, which intensified the aroma of the scented candles on display just under it. I knew I was in Accra, but I might as well have been in a New York boutique—or maybe an Italian one, given its owner's Italian-Ghanaian heritage.

The shop was called Lokko House, a warehouse-shop hybrid where designers and artists could gather and play with the many facets of fashion available to them at Kantamanto. Today's event was custom jackets, which could be adorned with patches or retooled in endless combinations according to one's tastes. I watched as other shoppers combed through the jacket offerings: labels like Supreme and Gucci were interspersed among the items, yet they weren't being gobbled up at the rate I'd expected. Brand names didn't matter to them as much as style, fit, and fabric—the real marker of good clothing—especially when what they cared about was the personalized alterations more than the original. The kids were smiling, dancing along to the DJ'd music, actually enjoying fashion. I am no designer, so I coyly asked one of the best-dressed men I have ever seen, a stylist named Amah Ayivi, if he might have some ideas for me; he was generous enough to dive in and design my custom-made jean jacket (it was a boy's Zara jacket) with Levi's and Wrangler patches in a grid on the back. If this is what we could mean by circular fashion—a way to wear denim in an infinite loop—then I'm 100 percent all in.

Ghana may be the last stop for secondhand clothing as it was origi-

Amah Ayivi and a sewer collaborating on my custom-made
jean jacket at the Lokko House pop-up.

nally worn, but the Instagrammers of Kantamanto are literally resuscitating them into something new and, if I may say so, dramatically improved. Unlike in the United States, where some secondhand clothing still has an air of dowdiness and misfortune (though businesses like The RealReal, ThredUp, and others are changing that rapidly, as well as new designers creating upcycled looks), Ghanaian designers are making secondhand cooler than new.

In addition to developing a domestic textile and clothing industry, secondhand clothing could be the raw material for a booming creative industry. Ghana could be well poised to start a revolution that turns secondhand clothes into treasure. "We live in a world of commerce, but that doesn't mean it has to be through the selling of goods—it could be through the selling of services like retailoring clothes and reconstructing clothes," says Anika, a German-Ghanaian woman I met at the store.

COVID-19 has revealed a silver lining for these incredible African designers. As imports of raw materials of secondhand clothes shrank, and workspaces closed, designers in Kenya, where imports were temporarily banned, began turning to new design and production projects that would keep the fashion industry close to home—in essence, the opposite of the Washington Consensus's aims. While it will take time for a domestic industry to mature, it's a valuable first step toward reclaiming some of that local autonomy Sammy told me about.

In addition to other positive media about the resale market and secondhand/upcycling, the role of Instagram influencers the world over is even more crucial for this growing population of shoppers and designers. If celebrities project an image that rewearing clothes is unacceptable, or vice versa, that will ripple out to the rest of the world, including these thousands of people who depend on rewearing at every level to survive. I'm hardly a celebrity on the red carpet, but walking down the sidewalks of Williamsburg in my newly designed jacket from Lokko House gave me a sense of pride and happiness unlike any garment I've ever worn. Knowing something so cool was crafted from something that might have otherwise burned in Kpone gratified me for more than an instant. Carefully choosing, buying, and wearing our clothes honors the work that people like Carl and Rima, Cesar and Danu, Dashi and Laura, Becky and Sammy do. If more of our purchases are truly loved, even if we eventually give them away, that means that they're less likely to end up as waste for someone someplace else, for someone like Vito, Major Fred, David, or Abena to deal with on our behalf.

Nine

Let the Makeover Begin:
Time for a New, New Deal

Knowing is not enough, we must apply.
Willing is not enough, we must do.

JOHANN WOLFGANG VON GOETHE

P hew. We have covered a lot of territory—around the world, more than once—unraveling an industry that is devastatingly destructive to our planet, its people, and our happiness. From this journey, we've seen how we, our country, and the world are stitched together, and how the tensions within and between these threads are tearing our natural, political, and economic systems apart at the seams.

So, that was fun, wasn't it? Let's go on our merry way.

Of course we're not finished. Now that we did all the work to understand how the system operates, we get to begin to solve its problems. Central to this is undoing the work of Bernays and his progeny. So here goes. This last chapter offers concrete ways to address the wrongs of the system through the different roles that we play in society—as a consumer, which is how Bernays would like us to see ourselves, and as a citizen, which I suggest is both our more powerful and more pleasurable role.

AS A CONSUMER

As consumers, we have two ways to affect the system: through our actual purchasing decisions, or "the power of the purse," and with our demands for better from brands we support—the power of our voice. Let's start with our individual purchases. While statistically insignificant on an individual level, how and what we buy can model better behavior for others, and those ripple effects are meaningful. And for influencers and celebrities reading this, your modeling behavior (perhaps literally) has an enormous impact on the rest of us (but more on that momentarily).

THE POWER OF YOUR PURSE
Organizing Our Closets

There are several very good resources for how to think about addressing one's wardrobe mindfully and sustainably. I particularly recommend *The Conscious Closet* by Elizabeth Cline and the website The Frontlash. These resources will guide and inform in detail, but I offer the following brief introduction.

To address one's wardrobe, one must of course see what's in it. And as Marie Kondo has told us a few times, this means getting rid of things. But before we do that, let's not forget the lessons from chapters 7 and 8, where we saw what actually happens to our discarded goods—cue the image of a burning Ghana landfill. Does that mean we can't purge our closets and spark joy anymore? Yes and no. To have a positive relationship with our clothing, we need to see what we actually have, and for many of us that means getting rid of the things that we don't actually wear and will likely never wear. But as you cleanse your closet, consider adding a seventh step to the six-rule method of KonMari: borrowing from another Japanese concept, *mottainai*, which means a regret for wasting a resource. This step encourages us to reflect on waste and will help us reduce, reuse, and ultimately enjoy our clothes.

Now we have a stress-free closet. The next time we go out in the world and consider buying something, we can ask ourselves the following questions so it stays that way. Think of this as a "shopper's checklist":

1. Does it spark joy?

Ask before you buy, not five years later, and you'll avoid being part of the statistic that women regularly wear only 20 percent of their wardrobes.

2. Does it fit me well?

In order to spark joy, a garment will probably need to fit, so if you can't find things off the rack, get to know a tailor; like the folks set up in Kantamanto, they can help make your off-the-rack, or out-of-the-box, clothes fit.

3. Does it feel good?

Just say no to clothes that don't feel good. The bonus? Quality natural fibers tend to feel better, so by going natural you do yourself a favor and you don't further contribute to microplastic pollution.

4. How is the garment constructed?

Look at the actual clothes with your eyes—the textile that was woven by someone, the seams stitched by someone else, the details added by someone else still. Turn the garment inside out and examine it. Are there threads coming out?

*Are the stitches uneven? Do they give when you tug on
them? If so, no matter how low the sticker price is, that
garment is not a good deal.*

5. Where will I wear it?

*I just informed you that most women wear only 20 percent
of their clothing, which means 80 percent doesn't have a
specific purpose in their lives. Whether it's formal wear or
everyday wear, choose items that you know serve your
lifestyle.*

6. This costs how much? Think cost-per-wear.

*I think of a purchase in terms of price per wear, rather than
the sticker price. I have spoken to many women, like
Tamara, who love the thrill of a sale, but what they buy in
those moments of euphoria winds up hanging in their
closets with the tags still on. If you're not actually going to
wear the thing to begin with, it's less of a steal for you, and
more of a steal for the store. Today, I see my clothes as an
investment. I may spend more per garment than I used to,
but I wear those pieces more and I feel better about myself
in my clothes, and I don't have to deal with the
psychological stress of clutter.*

7. Whom do I want to support?

*When you do decide to make a purchase, consider the type of
company you want to support. Look at who owns and
manages the company and how they market their goods. If
you want to live in a society that does not perpetuate
structural racism, be aware and make a commitment to
support diverse and specifically Black- or indigenous-owned
companies.*

Okay, so we've organized our closets, but there's more to how we can make a difference as consumers.

But What About Sustainable or Slow Fashion?

You will notice I did not start this section with the rallying cry to buy "sustainable fashion." I do very much appreciate the genuine efforts from brands to reduce their impact; we need more of these efforts and I know from running Zady how difficult it is. But buying "sustainable fashion" is not a panacea.

First, as I hope reading this book has revealed to you, there is really no such thing as "sustainable fashion." There is fashion that may have a relatively lower impact, but it is a matter of degrees, and the data collection has been so weak in this industry that it's often even difficult to conclude that a product labeled "sustainable fashion" actually has demonstrably lesser impact.

The truth is that the terms "sustainable fashion" or "slow fashion" have become so broad as to become almost meaningless, which is terrible for those companies that are actively trying to pursue genuine efforts in that space and the people seeking to spend consciously. As you know from chapter 6, there's much more marketing-speak about sustainability than implementation and evidence of real impact. This is just as true for entire companies that brand themselves as "radically transparent" as it is for fast-fashion companies that have a special "sustainable" capsule collection. There have been attempts to create consumer guides on this, but they themselves rely on public disclosures, which are so far insufficient to have any meaningful data to compare.

The second problem: So-called sustainable fashion perpetuates a notion that we can buy our way into sustainability, and that it's just a matter of purchasing this thing over that. This is not the case. The most sustainable thing is to not buy the thing at all. This doesn't mean that you shouldn't buy anything; it means cutting out the things that don't spark joy to begin with, and regaining control over our closets and our lives.

Finally, it is challenging that items marketed as "sustainable fashion" are out of financial reach for many people. While clothing should never be as cheap as it is on the very lowest end (a T-shirt that costs $2, given all of the steps we went through, means that someone's well-being, in addition to the planet's, is being compromised), what we can do is to wear our things more and search for higher quality, which can be found at a variety of price points, especially when we take into consideration cost per wear.

To be clear, this is not to belittle the significant efforts many brands are making. The point is, at this stage there is no data-driven, systemic way to judge brands that are marketed as sustainable.

Explore Resale (Carefully)

Take clues from the fashionistas in Ghana and shop secondhand. Secondhand is fast becoming not only acceptable but a bragging right. Head to sites like ThredUp or The RealReal in the United States, Depop in the United Kingdom, Vestiaire Collective in France, or Kiabza in India—they are part of the fast-growing digital secondhand market, which (and this is according to a report that ThredUp commissioned, so take it with a small grain of salt) is expected to overtake fast fashion and be 50 percent bigger than it by 2029.

I have sold pieces on The RealReal and have bought great pieces there that I would not otherwise be able to afford. But resale isn't the entire answer either. There is some evidence as I describe below that the marketing around circularity may create even more consumption: People buy something new and think, "I'll be doing a good thing because I'll be selling it into the secondhand market," and those who buy from the secondhand market think, "Well, I didn't buy anything new." But those mindsets enable each other and the consumption machine. So, buying used is better than buying new, but not buying is by far the best of all.

Develop a Healthy Relationship with Social Media

To get ourselves to buy less, we've got to start with cutting out all of those "Buy me, buy me," "You know you want me," and "See this thing? If you buy it, you will have no problems and be perpetually happy, and look fabulous while you live your dream" messages. For many of us, ground zero exposure to those teases is social media, specifically Instagram. Instagram is an advertisement-based platform. My Instagram feed has one ad for every three non-ads that I follow, not to mention the influencers and celebrities who promote products in their posts. Unfortunately, we are not allowed to turn the ads off in our settings.

The low-tech fix? Start by editing down whom you follow, so that at least the non-Instagram ads are not just influencers' #ads. I try to cut out accounts that give me FOMO or that just don't make me feel good. If I am going to spend money, I want it to be a positive experience, not because I feel inadequate. And then, of course, I periodically just delete the app altogether and enjoy being present with my family and doing things like exercising, reading a book, seeing a friend, or watching a good movie. You can do the same with your email account—just click unsubscribe.

Remember what happened when cigarette ads were banned? Fewer people smoked. We apply the same principle to shopping addictions—less exposure to ads and other messages of consumption, fewer mindless purchases.

The Circularity Myth

Sustainability-minded readers will have heard *a lot* of talk about circularity. H&M has committed to becoming 100 percent circular and has a page dedicated to the concept; Gap has its own 2020 Circular Fashion Commitment.

Sounds great, I know, but I have a significant bone to pick with this supposed "solution." It reminds me of something I picked up in law school—the idea of "moral hazard." For example, being a more reckless driver because you are covered by insurance. The same general idea can be applied with circularity. If we believe that if we just bring our clothes back to H&M and they will get magically turned into new clothes, the less trouble we have buying more stuff. There has in fact been some proof of this mindset: Researchers have found that the presence of a recycling bin in a bathroom actually correlated to more paper usage. Many retailers not only put their recycling bin in the store to encourage people to bring clothes back, but they also financially incentivize you to do so. It's as if the recycling bin in the bathroom gave you money to put paper there, which you could use to buy a new roll of paper towels to bring home with you, in the paper towel store conveniently located next to the bathroom. But in reality, those take-back bins at H&M et al. follow a very similar path to the clothing dropped off at the Salvation Army. So the long and short of it is, while company investment in recycling material is very important, circularity as a marketing tool is the moral hazard of fashion. Don't fall for it.

A Special Note for Influencers/Celebrities

Hello, influencer/celebrity, this message is for you. I know and totally appreciate that part of your current business model is to sell things. But the world is getting flatter, and the more you promote buying stuff or wearing new things every time you are photographed, the more the entire world gets on board with the idea that we need new stuff, and can only wear things once. And the more both messages are tweeted, reposted, and hearted on the Internet, the faster the planet is destroyed and the more people are put at risk.

You are powerful. Your actions have massive global implications. You are in control of the faucet of consumerism. So, please, use that power for good. Consider evolving your own brand into one that relies less on others buying stuff that is ultimately destructive to the world. If you show yourself wearing clothes more than once, that becomes the aspiration for the rest of us. If you don't, well, that becomes our aspiration, too. Instead of stuff, consider promoting services: yoga classes (not yoga pants), or meditation (not pretty beads), or cooking lessons (not trendy cookware)—you know, experiences that will bring people the happiness they crave.

Our clothes can be things that we love and that make us feel good. Our closets can be stress free—a place to see and enjoy the armor we'll use for the day. In the next section, I will share not only how our clothes can be a great purchase for us and can contribute to slowing down the system, but also how we can use our voices to regulate the industry and move it to one that releases only clean water and air with workers that are given the tools to thrive.

THE POWER OF YOUR VOICE

So you've adopted this beautiful, conscious, healthy relationship to buying, and that's amazing! Well done. Our personal choices can begin to slow down an industry that has gone off the rails, turning four seasons a year into nonstop-new. But it's only through advocacy that we can get companies to change their ways: decreasing their carbon use, managing their chemical use, paying their workers a decent wage, and having companies that do not perpetuate structural racism—all the things we talked about earlier in the book.

The interesting thing about the clothing industry is that its profits are highly consolidated. The top twenty fashion companies hold 97 percent of the profit of the industry. Creating a shift in just these companies, we can achieve significant gains globally. Believe it or not, executives read the emails they receive and listen to the chatter on social media. We, the consuming public, create the demand that companies fill. Demand that companies draw down their carbon use, manage their chemicals, pay their workers a decent wage, and be representative and inclusive and they will do that. If we do not demand this, change will not happen.

This is central to our work at New Standard Institute, to support the industry by keeping it accountable. We need the large fashion companies—and all industries—to measure their environmental and social footprints, to disclose this measurement, to publicly set targets, and to disclose how those targets are being achieved. Making progress is not rocket science, it's just a matter of corporate will. You can play a central role in creating that will by using your voice to ask (or demand) companies for this kind of accountability. Join the NSI community at @NSIFashion2030 on social media or www.newstandardinstitute.org; there you can sign our petition demanding this accountability. We also provide the most up-to-date information on the industry and avenues for making your voice heard. As I said, only if they hear from you—which is the only way they'll know you have an opinion—will they move in this direction. If we do not say anything, we can expect much of the same: vague marketing claims to get us to buy even more stuff (but this time labeled "sustainable fashion"), but nothing that actually solves the enormous and very real problems the industry and the planet face.

And, as we have seen throughout the journey of our clothes, all too often data on the environmental and social impact of this industry is very hard to come by, which makes judging progress all but impossible. We only manage what we measure. The leading players in the industry have to step up and fund the filling in of these data gaps. Funding has to be completely separate from any role in the research that is produced, and the research must be done by the actual experts in the field, not the same consulting firms that are hired by those companies to continue to grow. NSI is the independent gateway for existing information; we have brought together the leading researchers from around the world. They are the ones to lead the way in filling in the gaps and ensuring accountability.

A Special Note for Finance
(and All People with Bank Accounts)

Finance has a potentially huge role to play in addressing the global challenges covered in this book. Finance—the shareholders who actually own the companies—could have enormous positive influence by insisting that environmental and social targets within a company are established and reached.

The good news is that there is a growing interest in using finance for good. An entire booming subfield within finance called ESG (environmental, social, and corporate governance) has been developed to advance these goals. Banks from BlackRock to Wells Fargo to UBS have all gotten in on the game. In recent years $30 trillion (yup, with a "t") has been put to ESG investment; more than a quarter of publicly listed companies around the world are now ESG measured and rated.

But ESG risks becoming meaningless and greenwashed if its metrics do not actually capture the social and environmental impact of a given company. As Sasja Beslik, head of sustainable finance development at J. Safra Sarasin Ltd, describes, the problem is that "current ESG data puts the spotlight on what is available, rather than what is most important." For example, companies often list greenhouse gas emissions of their own operations as the metric for their climate work, but as we saw in chapters 1 and 2, the hotspot for climate impact in fashion is not the energy efficiency of the lightbulbs at Zara headquarters, it's the enormous, fuel-devouring supply chain. The exact same demands that consumers should make that I discussed above apply to financial companies, too; financial institutions must demand companies measure their environmental and social footprints, disclose this measurement, publicly set targets, and disclose how those targets are being achieved. Anything short of that would not capture the actual impact of a company and would therefore be meaningless. Any "sustainable" investment product sold to a customer with wishy-washy metrics is just as problematic as a fast-fashion company selling a vaguely worded "green" T-shirt.

As a consumer of financial products, you have a voice with your bank as well. Ask them if they are demanding that companies provide meaningful disclosures and work toward material reduction targets. You may think that your voice is a drop in the ocean, but it's this chatter that starts much bigger conversations in banks and at companies, and that actually leads to meaningful progress.

AS A CITIZEN

In discussing how we solve the challenges of the clothing industry, we cannot help but bump into many of the central issues of our time. By seeing where fiber is produced, we see that the organic standard that I once believed was synonymous with "sustainable" is actually not. We see that to solve climate change, we also have to address income inequality, since it's going to be difficult to persuade people who are financially stressed to prioritize climate change when their basic needs are on the line. In looking at the origin of cotton and the lack of union representation in today's distribution facilities, we see that the clothing industry itself drove enslavement and the appropriation of indigenous land and has continued to perpetuate structural racism. By exploring where our clothing goes when we get rid of it, we uncover how developing countries are put at a disadvantage through trade deals. By seeing where our clothes are made, we uncover international trade policy and how hyperglobalization pits workers globally against one another. We see how corporate executives have made workers' voices weaker, and how that has contributed to income inequality in this country. We have learned that there has been an enormous investment in getting us, citizens, to see ourselves as docile buying machines, which social media has only exacerbated, rather than powerful stakeholders in our democracy. Combined, we have seen how extreme neoliberal capitalism has failed us and the planet. My sense of citizenship has been forever altered. I will not just be a consumer, passively consuming while things fall apart. I will be an active citizen engaging in community and with government to build a just and thriving society.

The following are some of the policy areas that have come up in the journey of our jeans. It is far from an exhaustive list, but I hope it can serve as a starting point for reflection and action.

Agricultural Policy

Climate change can be tackled in two ways: (1) by limiting carbon emissions, i.e., stop burning fossil fuels, and (2) by sequestering more carbon from the atmosphere. Soil may play a significant role in carbon sequestration. Before this book, I assumed that "organic" and "sustainable" were interchangeable terms in farming. I think this is what most people still believe. It is not exactly true. The organic standard has been a significant development, but agricultural policy going forward needs to be based on maximizing long-term output and soil health, and more research needs to be done to better understand both how this is achieved and its connection to carbon sequestration and with it climate change mitigation.

Climate Policy

It goes without saying, but I'll say it: We need an aggressive climate policy. Moreover, as we explored in chapter 2, our climate footprint reaches far beyond our own borders, and yet when climate policy is examined, this fact is not always considered. Our climate policy should of course be grounded in drawing down carbon emissions domestically, but we must also consider how we can structure trade policy to ensure that we are not just displacing those emissions by producing our goods—and those emissions—somewhere else. If we reduce emissions domestically only to increase them outside our borders, we haven't really done anything. That's just moving the deck chairs on the *Titanic*.

Trade Policy

The United States is the largest market, and because of this we have a lot of leverage. We can use this to our and the planet's advantage. We can make recognition and enforcement of core labor rights a precondition to trade, which would help ensure that our clothing is not just made in the places with the most lax labor laws. We can also stop the race to the bottom of cheap and dirty energy use that we saw in chapter 2 by imposing a border carbon adjustment, so that if a brand imports clothing that uses carbon-intensive processes, it would be charged a fee. And we can stop being a roadblock for potential progress in the developing world by, as a starting point, not requiring other countries to remove tariffs on the importation of our discarded goods as a prerequisite for beneficial trade, as we saw was the case when Rwanda attempted to use some protectionist measures—just as the United States, Europe, and China have done in the past—to develop its own clothing manufacturing sector.

Addressing Our Past and the Structural Inequalities of the Present in the United States

Through the journey of our jeans we have seen how Native Americans were driven off their land and how the cotton industry fueled enslavement. And in chapter 5 we got a glimpse of how racist laws and then racist views have led to structural racism today. Things like increasing the federal minimum wage and ensuring union representation are important starting points to narrowing the race-wage gap. There is much, much more to be done on this front, but the fashion industry would be wise to recognize the central role that our clothing has played to create the conditions today, and because of this, it should take a leadership role in addressing them. It can start by ensuring that the leadership of its own companies is both representative and inclusive.

From Winner Takes All to a Prosperous Society

In the life of our jeans we have come across a lot of income disparity. Farmers may struggle, garment workers struggle, former garment workers in this country struggle, distribution workers struggle, the people left selling the lowest-quality secondhand goods struggle, and then a very few people profit enormously. If we can have a system in which those struggling people could also thrive, wouldn't we want that?

I support policies that provide worker protections. I want to live in a society with basic social protections, where birth alone does not almost entirely determine outcome. And I'd like the jeans or shirt or dress that I wear and enjoy not to be part of a system that is a significant contributor to climate change and leaves the people who have touched it with enormous economic stress. I will not forget the faces and stories of the people I met while writing this book.

Neoliberalism and Deaths of Despair

Exploring the history of the clothing industry and how it treated workers on its race to the bottom—turning garment and distribution workers into zombified machines, while the founders of those companies march up the *Forbes* billionaire ladder—has made me seriously question neoliberalism. Even when GDP has boomed, the lives of the workers at the bottom remain devastatingly difficult. As we touched on in chapter

5, in the United States since 2005 the incidence of "deaths of despair" is growing. And they have made the United States the only high-income country in which life expectancy is moving in the wrong direction. There are of course other factors, including a lack of access to healthcare and access to opioids, but having spoken to workers whose jobs are mind-numbing, where their salaries make their livelihoods and lives precarious, I see that the market, left exclusively to its own devices, is not proving to be the perfect solution.

Again, this goes back to what we measure, and by extension what we value. So far, many economists and people in media have put an almost singular focus on GDP as the standard bearer of the health of our society, but this has proven to be misguided. GDP measures the size of a country's economy—in other words, the size of the market. But it does not measure the nation's welfare. As Amit Kapoor and Bibek Debroy stated in *Harvard Business Review*, "When our measures of development go beyond an inimical fixation towards higher production [GDP], our policy interventions will become more aligned with the aspects of life that citizens truly value, and society will be better served." Measuring things like access to housing, healthcare, and childcare, and the gender wage gap, the race wage gap, and the wage gap between executives and the average worker in a company, can all be metrics to help us more truly understand the welfare of our society.

The Future of Work and Industrial Policy

We saw how globalization as it has been structured caused job loss in the apparel manufacturing sector in the United States. And we have seen how machines are getting smarter and more agile, replacing workers both in the cut-and-sew factories and in distribution centers. We also know that if we succeed in slowing down fashion consumption, we could, in theory, see a contraction of the industry—although, to be very clear, this is highly theoretical as (1) we're all still going to get dressed; (2) the global population is still increasing; and, COVID-19 aside, (3) the world is getting richer and consuming more. All that is to say, I'm not losing too much sleep over the fashion industry collapsing, even with the sharp contraction in clothing sales brought on by COVID-19. But we do need to think more about what industries we do want to see.

Markets cannot solve this for us. Instead, government officials must design industrial policy for the world we live in, creating incentives for growth in certain sectors. Investment in renewable energy, for example, will help create jobs and have those jobs contribute to a sustainable world. We saw how government support combined with market forces allowed both France in its time and China today to dominate apparel production, and the government there is doing the same with its Made in China 2025 strategy today. The United States and other Western countries would be wise to follow suit or seriously risk being left behind. What industries do we want that come with well-paying jobs in which we can build national expertise? Green energy, low-impact materials, green infrastructure.

And for those jobs that are physically taxing and mind numbing? Well, I think that is a conversation we need to have in society, not just fashion. The writer Johann Hari explored the leading causes of depression in his book *Lost Connections*; the first one he lists is "disconnection from meaningful work." In the journey of our jeans, we've seen a great deal of this type of not-meaningful work. Hari quoted researcher Michael Marmot: "Disempowerment is at the heart of poor health."

Instead of relying on individual companies to take the high road, which we have seen just does not happen at scale, we must elect politicians who support policies that empower workers by improving workplace conditions, strengthening workers' voices, and addressing the runaway costs of childcare and healthcare that have left Becky, Laura, Sam, and millions of Americans out on a limb. And we need industrial policy that looks beyond just creating more jobs—those jobs need to be good jobs, in Dani Rodrik's terms, so that all those people throughout the supply chain of our jeans have a quality standard of living. Here's what I mean by that:

Improve Workplace Conditions
and Workers' Voice

- Back legislation that supports fairer scheduling practices.
- Have workers on corporate boards. This is a model we covered in chapter 3 that has been effectively implemented in Germany to ensure that workers have a voice in the direction of the company.
- Fix how executives are incentivized. The majority of executive compensation is through stock, as we covered, which induces executives at companies like Amazon to focus exclusively on short-term shareholder value. If we limit the sale of stock for a period of time, it can ensure that executives look at longer-term investment, which can lead to more jobs and other stakeholders—the workers.
- End antiworker laws, misnamed "right to work laws," which prohibit unions and employers from agreeing that any employee who benefits from a union contract should have to pay dues to support the union. These state laws deprive union treasuries of funds needed to effectively represent workers.
- Get rid of provision 8(c) of the Taft-Hartley Act, which permits employer interference like encouraging worker attendance at antiunion meetings and restricting reasonable union access to the workplace. Laura shared an experience of this back in chapter 5, when she explained how she felt talking about unions was stifled at Amazon.

Develop Industrial Policy

- As we have seen throughout this book from the Sun King in France, the Six Priorities of China, and the New Deal in the

United States to the modern economies of China, Germany, and Japan, growth has been fueled in part by the support of government. Yet today America spends half as much on federal research and development as we did in the 1980s. We can increase investment in research, ensure that production that comes out of that research takes place domestically, and pair this with investment in education—including apprenticeship programs—and use this to fuel Rodrik's good jobs economy.

Address Runaway Costs, So That Pay Can Lead to Middle-class Life Through:

- Affordable childcare
- Universal healthcare
- Increased supply of affordable housing
- Student loan relief

NO ONE IS GOING TO DO THIS FOR US. WE ARE ALL RESPONSIBLE

So many of the topics that have come up in the journey of our clothes end up seeming larger-than-life—economics, the environment, chemical standards, disparities based on race, global trade. When I embarked on my research and reporting—I should note this took several years—I had a vague notion that the government was regulating these areas, or at least that environmental organizations or other nonprofits were taking care of it. Besides, I'm not an economist or a psychologist or a chemist. What could I do other than try to address issues within my own community?

But the more I peel back the layers of the fashion industry, the more I see that things don't just happen behind the scenes. It is, more often than not, the contrary. Companies put out corporate social responsibility reports most often to *signal* they are doing something rather than to actually address core challenges. Meanwhile, they send lobbyists to create laws that benefit themselves, at a significant cost to us. Environmental organizations may have blind spots around impact outside the countries in which they operate—which is certainly the case in the fashion industry—and they are waiting on us, the voting public, to get involved. Some have also found themselves compromised by their corporate donors, which makes messier topics, like rampant consumerism, untouchable. Meanwhile, we've been fed more fashion to distract us from the governance that is going on without us.

We have to break through the work of Bernays and his followers and step away from our role as consumers and find our way back to citizenship. We play the critical role in being active citizens. That can take a variety of forms—marching, calling, emailing, tweeting. Certainly, voting. It means asking questions at a store, writing enquiring emails to companies, and participating in calls to action. While I was writing this book, the fabric of our society has shown how it has begun to unravel, but it has also shown the power of people taking to the streets, the beauty of a truly diverse movement of advocates energized and ready to address the fundamental problems in our society.

To be very honest, advocacy does not come naturally to me. I don't like to rock the boat. But now I know that if I do not become more active—if we do not become more active citizens—change is not coming. This is something I have learned through writing this book, and I have become more energized by the passion of the advocacy community that has come forward while this book has been in process.

The upside of citizen engagement? My journey has given my life another layer of meaning. Through it I am finding lifelong friends and a feeling of community, many of the things experts have pointed to as critical building blocks of true happiness. By working with others to be a part of addressing the great challenges of our time, I am happier. And I didn't even need retail therapy as a Band-Aid to get me there.

WHAT WE CAN CREATE:
A VISION FOR THE FUTURE

When we focus too much attention on what is wrong, we have a hard time seeing a way forward. So in that spirit, I want to conclude with a vision of what we could build if only we stand up and invest in ourselves, our community, and our society.

In 2030, our closets are tidy, rather than overstuffed. We love everything in them and wear our clothes deliberately and often. The clothing itself is made of remanufactured material, originally grown on a farm using regenerative practices. The soil from which it grows is teeming with minerals, organic matter, and microorganisms, which are helping absorb carbon from the atmosphere. The farmer and her workers are compensated sufficiently for doing the work of growing the plants that become my jeans.

That cotton is woven at a mill that runs on solar energy, and the chemical usage is overseen by an independent body led by toxicologists. The fashion brands that work with the textile mills have codes of conduct in place overseeing these processes that they actually enforce, and work closely with the textile mill to see that what they are being paid can cover the costs to ensure this safe management. Kids jump and splash and swim in the rivers near the factories because the water is treated properly. Fashion brands publicly disclose their environmental footprint and have achieved the public targets that they had set out for themselves.

Our jeans get expertly sewn together by workers who take satisfaction from their challenging jobs, contributing something beautiful that will make another person feel great for a long time. Those workers responsible for making my jeans are paid a wage that allows them to live fulfilled and healthy lives like I do. It means my jeans cost a bit more, but we're okay with that because we love the jeans and know that we will enjoy wearing them for some time.

Our government has created an effective industrial policy, which has created meaningful, well-paying jobs and allows me to visit family and explore the country in a cool, high-speed train.

We know we will love our jeans because we have wrested control of our own attention and removed the noise in our inboxes and on our social media channels that had distracted us from our true needs and desires. And we bought them not for the dopamine hit we need to fill other holes in our lives, but from an aware and informed mindset.

That garment is packaged and delivered to us by workers who are unionized and have a relationship with management to ensure that their voices are heard, and they understand the environment in which the industry is operating. Their wage affords their kids the same opportunities for success that the rest of us have.

The brand we have purchased from has a representative and inclusive management and board, to lay the foundation for a more inclusive company and to be part of dismantling a racist system.

But most of the time, we're not shopping for clothes at all. We're just wearing them while out having fun with friends or family, or engaged in local politics. We are less anxious and sleep well at night. When we read articles by the media, the reports are fact-checked and have robust systems in place to ensure that the information we are reading is accurate, including reporting on the fashion industry.

When the jeans we bought no longer suit us, we don't throw them out unless they're unwearable—and then, we know that they'll decompose in a well-managed compost. We have had a referendum on race and structurally addressed racism so race will no longer be the most significant factor in who lives by our toxic waste sites. And besides, those waste sites will be well managed, with long-term regulations in place. Otherwise, when we do give our clothes away, we give them away as a donation to a facility that can then sell them as a garment of quality. Meanwhile, in New York, Accra, and London, teams of designers develop exciting new designs from this existing material.

The oceans teem with fish, not plastic; our forests burst with trees and animals. The fashion industry is not dismissed as girly, but seen for the genuine, significantly positive impact that it has.

———

The global fashion industry and the global economy are not a force of nature unable to be tamed. They were designed and can be redesigned just like our jeans. The choice is ours.

Acknowledgments

This book was several years in the making, and I owe thanks to many people. Back at Zady, I had wanted to create the time to travel the supply chain of Big Fashion and do all the research to fully connect the dots that I started to see while there. The idea was parked very much in the back of my mind, until I caught up with my friend Greg Behrman, who replanted the idea, encouraged it, and was kind enough to connect me to The Cheney Agency, where I met Alice Whitwham, who would become my agent. Alice believed in the concept and its urgency long before "sustainability in fashion" was trendy. This book would not have happened without her support, guidance, and friendship. I am also grateful to the brilliant Eve MacSweeney, who worked with us on the proposal and helped me find my voice.

I relied on the stories and expertise of so many people working in many different domains that touch on the story of our jeans, and they were all generous in sharing their time and knowledge. In Texas, Cesar Veramontes opened up the world of denim in the United States, while Kent Khal and Carl Pepper introduced me to cotton farming and made this city girl feel very much at home. In China, Charles Wang opened his factory to me and was an equally generous guide in Guangdong. I will never forget our tea; it was a real highlight. Charles connected us with Dashi, who is an inspiration. In Sri Lanka, Ashila and Danu took the time to share their stories and introduce us to other garment workers in Colombo. Melani Gunathilaka served not only as a helpful volunteer interpreter, but also shared her important perspective. In Bangladesh, I am grateful for the help of Professor Shahidur Raman, Rob Wayss, the team at the ILO, and, of course, Rima. And, in Ghana, I'm so grateful to Liz Ricketts, who so generously took me into the extended family that she has developed in Accra. Part of that family included Abena, Stefania Manfreda, and David Adams, who is a brilliant mind. I'm looking forward to seeing where that mind takes him. As Stefania said, there is a lot of Wakanda energy in Accra; I am most grateful for everyone who shared that energy with me. Back in New York, the team at DSNY gave me renewed faith in city government. Much gratitude to Vito and Major Keith Mellis for letting me peek into the life of my garbage. And at the Salvation Army, many thanks to Fred Muhs for showing me the fascinating world of our donated goods. For the people I consulted with who cannot be named, I would like them to know how much I appreciate them for sharing their stories with me. I only hope I have done them justice.

Back at my desk, I reached out to the leading experts in their fields and am grateful that all took the time to share their knowledge with me. Dr. Christoph Meinrenken is my guru for all things climate change. That he still manages to have a sense of humor doing his work is, well, impressive. Dr. Linda Greer helped me understand what I saw when I was in China. Dr. Greer has also played a hugely significant role in the fashion industry; as the unofficial chief scientist, she combines blunt talk with an insatiable charm that everyone wants to be around. If the industry does measurably improve, we will all have Linda to thank. I want to also acknowledge the contributions of Drs. Sven Beckert, Dani Rodrik, Darby Saxbe, Wendy Wood, Tim Kasser, Bob Yokelson, and Keshav Kranthi, as well as the contribution of Peter Anderson.

If anything has moved you in this book, it is with enormous thanks to Jennifer Kurdyla, who was instrumental in taking my rather dry research and making the text into what we hope is a moving narrative. Jennifer not only drew out the story, she has been an incredible thought partner as together we waded through stories, industries, and research. She is very much a silent partner in this book, and working with her has been a career highlight for me.

My editor at Portfolio/Penguin, Merry Sun, who believed in the topic and pruned and tended magically to the text, finding just the right balance between data and narrative to make everything come alive. She has understood from the beginning that fashion is far more than a silly industry, and has helped me find the words to make sure that that point is made as strongly as possible. Thank you, thank you, thank you. And my gratitude to the team at Penguin/Portfolio, who have gotten the book in shape and positioned it so thoughtfully.

Allison Deger was fact-checker extraordinaire. Fact-checking plays an unspoken but critical role in society; I was lucky to work with her. And Jane Cavolina, who turned the mess that was my citations into a resource that I hope will be useful for others.

My colleagues at New Standard Institute: George MacPherson, who has been patient every time I disappeared to write, and Mathew Swenson, who has championed the idea of all of us putting a mirror to ourselves and the industry. And my deepest thanks to and gratitude for Alejandra Pollak, who combines great research skills and delightful charm, and did much of the initial research for the book and was the most fun travel companion. Your generosity of spirit is something I deeply admire and for which I will remain forever grateful. You all renew my faith in humanity, and I'm so grateful to get to work with you.

I hadn't realized how researching and writing a book can be a rather lonely endeavor, but my friends were never more than a text message away and created the support I needed to continue. The research shows that connection drives meaning and happiness, and my friends have certainly given me that.

Giselle Yameago has helped take care of my daughter so that I could write. Her delicious food kept me nourished, and my husband and I both got the peace of mind that our child was in the very best hands. Her contribution cannot be overstated; without her this book would not have been written.

And thank you to my parents, Keith and Valda, who have always encouraged me to do the work that I find most meaningful. What an enormous gift. My dad believed in me before I believed in myself, and had a work ethic, curiosity, and sense of adventure that I attempted to embody in this book. He was so excited about my writing, and I miss him every day so very much. Mom, your charm and dedication to our family and your community sets a bar I will never attain. Jess, Deb, and I are the luckiest daughters in the world to have you as our mother. You are simply the best. And I am also so very, very lucky to have big sisters who taught me everything, including to care about the world and our role in it. I am immeasurably proud to be a Kaye sister.

Finally, my husband, Stephane, who took care of our Leontine as an infant to allow me to run around the world doing research and who has always encouraged my work. I am extraordinarily grateful to have such a copilot in life. I love you.

Notes

INTRODUCTION

xi **doubling in the fifteen years:** Nathalie Remy, Eveline Speelman, and Steven Swartz, "Style That's Sustainable: A New Fast-Fashion Formula," McKinsey & Company, October 20, 2016, https://www.mckinsey.com/business-functions/sustainability our-insights/style-thats-sustainable-a-new-fast-fashion-formula.

xi **100 to 150 billion:** Randolph Kirchain, Elsa Olivetti, T. Reed Miller, and Suzanne Greene, "Sustainable Apparel Materials," Materials Systems Laboratory, Massachusetts Institute of Technology, September 22, 2015, https://matteroftrust.org /wp-content/uploads/2015/10/SustainableApparelMaterials.pdf.

xv **Today, 1.25 billion:** "Infographic: Data from the Denim Industry," Fashion-United.uk, September 26, 2016, https://fashionunited.uk/news/business/infographic -data-from-the-denim-industry/2016092621896.

xv **average American woman:** *ShopSmart*, "Jeaneology: ShopSmart Poll Finds Women Own 7 Pairs of Jeans, Only Wear 4," September 2010, available at https://www .prnewswire.com/news-releases/jeaneology-shopsmart-poll-finds-women -own-7-pairs-of-jeans-only-wear-4-98274009.html.

xv **$2.5 trillion industry:** Imran Ahed et al., "The State of Fashion 2019," McKinsey & Company, https://www.mckinsey.com/~/media/McKinsey/Industries/Retail/Our %20Insights/The%20influence%20of%20woke%20consumers%20on %20fashion/The-State-of-Fashion-2019.ashx.

xv **destructive impact on our environment:** Achim Berg and Karl-Hendrik Magnu, "Fashion on Climate," McKinsey & Company, 2020, https://www.mckinsey.com /~/media/McKinsey/Industries/Retail/Our%20Insights/Fashion%20on%20climate /Fashion-on-climate-Full-report.pdf.

xvi **"Too often we prefer":** Sven Beckert, *Empire of Cotton: A Global History* (New York: Knopf, 2014), xviii.

xvii **That story begins in India:** Organic Cotton, "The History of Cotton Production," organiccotton.org, https://www.organiccotton.org/oc/Cotton-general/World-market /History-of-cotton.php.

xvii **workers in the port city of Dongri:** "History of Dungaree Fabric," HistoryofJeans .com, http://www.historyofjeans.com/jeans-history/history-of-dungaree-fabric/.

xvii **Cotton in any form:** When European merchants landed in Dongri and elsewhere in Southeast Asia, their durable work pants, which were Anglicized into "dungarees,"

caught on among sailors who needed something that could be worn wet or dry, and
were stiff enough to be rolled up for activities like deck cleaning; it was also used as
a fabric for things like ship sails. Eventually, the French wanted to get in on the game
and came up with a copycat material. Their version of "dungarees" came from
Nîmes—"de Nîmes" in French became denim to us. You say denim, I say jeans, and
while today there's no meaningful difference between the two words, we can thank
the French again for making our vocabulary complicated: It is believed that "jeans"
originates from the French word Gênes, after the Italian port city of Genoa, where
the Italian sailors pulled off the hard-cotton-pants look with enviable panache.

xviii **cotton textiles accounted for:** Beckert, *Empire of Cotton*, 33.

xviii **there's the labor of picking:** "King Cotton," Utah Social Studies, Agriculture in
the Classroom, Utah State University Cooperative Extension, https://cals.arizona
.edu/fps/sites/cals.arizona.edu.fps/files/education/king_cotton.pdf.

xviii **the institution of slavery expanded:** Ronald Bailey, "The Other Side of Slavery:
Black Labor, Cotton, and Textile Industrialization in Great Britain and the United
States," *Agricultural History* 68, no. 2 (Spring 1994): 35–50, https://www.jstor.org
/stable/3744401?seq=1

xix **In 1850, 3.2 million people:** Henry Louis Gates, Jr., "What Was the Second Middle Passage," *The African Americans: Many Rivers to Cross*, PBS, https://www.pbs.org
/wnet/african-americans-many-rivers-to-cross/history/what-was-the-2nd-middle
-passage.

xix **formal workplace management:** Caitlin Rosenthal, "Plantations Practiced Modern Management," *Harvard Business Review*, September 2013, https://hbr.org/2013
/09/plantations-practiced-modern-management.

xx **the economy in its current form:** "2020 Edelman Trust Barometer Reveals Growing
Sense of Inequality Is Undermining Trust in Institutions," Edelman.com, January 19,
2020, https://www.edelman.com/news-awards/2020-edelman-trust-barometer.

CHAPTER 1: GROWTH MENTALITY:
COTTON FARMING IN TEXAS

1 **United States comes in third worldwide:** "Cotton: World Markets and Trade,"
World Production, Markets, and Trade Report, October 9, 2020, United States
Department of Agriculture, Foreign Agriculture Service, https://www.fas.usda
.gov/data/cotton-world-markets-and-trade.

1 **produces 40 to 50 percent:** "Texas Cotton: 'The New King,'" Lubbock Cotton
Exchange, https://www.lubbockcottonexchange.com.

1 **9.6 billion pounds are estimated:** "Cotton: World Markets and Trade," World
Production, Markets, and Trade Report, October 9, 2020, United States Department of Agriculture, Foreign Agriculture Service, https://www.fas.usda.gov/data
/cotton-world-markets-and-trade.

2 **grown in eighty different countries:** Terry Townsend, "Cotton in the World Economy," Cotton Analytics, July 19, 2018, http://cottonanalytics.com/cotton-in-the
-world-economy/.

2 **58.5 billion pounds:** "Cotton and Wool Yearbook," United States Department of Agriculture, Economic Research Service, November 21, 2019, https://www.ers .usda.gov/data-products/cotton-wool-and-textile-data/cotton-and-wool -yearbook/#World%20Cotton%20Supply%20and%20Demand.

4 **the Cooperative was responsible for 2 to 3 percent:** Interview with Kelly Pepper of Texas Organic Cotton Marketing Cooperative, October 1, 2020.

5 **the USDA organic standard:** United States Department of Agriculture, Agricultural Marketing Service, Grades and Standards: Organic, https://www.ams.usda .gov/grades-standards/organic-standards.

6 **$1.00 per pound:** "Cotton Prices—45-Year Historical Chart," Macrotrends.net, https://www.macrotrends.net/2533/cotton-prices-historical-chart-data.

6 **tiny portion of all cotton produced:** "Get the Facts About Organic Cotton," Organic Trade Association, November 18, 2019, https://ota.com/advocacy/fiber-and -textiles/get-facts-about-organic-cotton.

6 **suicide rate for male farmers:** *Morbidity and Mortality Weekly Report* 67, no. 45 (February 22, 2019): 186, https://www.cdc.gov/mmwr/volumes/68/wr/mm6807a7 .htm; Matt Perdue, "A Deeper Look at the CDC Findings on Farm Suicides," National Farmers Union, November 27, 2018, https://nfu.org/2018/11/27/cdc-study -clarifies-data-on-farm-stress/.

7 **transfer to cotton farmers of $2.1 billion annually:** "US Cotton Subsidies Insulate Producers from Economic Loss," Gro Intelligence, June 6, 2018, https://gro -intelligence.com/insights/articles/us-cotton-subsidies.

7 **institute policies in their favor:** The legitimacy of this tactic is coming under fire of late. In fact Brazil brought a case to the WTO about this—that the government subsidies drive up prices, which makes it that much harder for poor farmers in other countries to compete. "Critics like aid group Oxfam have long argued that U.S. cotton subsidies make it impossible for West African farmers to compete by depressing world prices by up to 14 percent." See Missy Ryan, "U.S. Cotton Could Suffer If It Loses Subsidy Support," Reuters, December 20, 2007, https://www .reuters.com/article/us-usa-cotton-wto/u-s-cotton-could-suffer-if-it-loses-subsidy -support-idUSN2060217920071220.

8 **by forced labor, including in Xinjiang:** Ana Nicolaci da Costa, "Xinjiang Cotton Sparks Concern over 'Forced Labour' Claims," BBC News, November 13, 2019, https://www.bbc.com/news/business-50312010; "Forced Labour in China Presents Dilemmas for Fashion Brands," *The Economist*, August 20, 2020, https://www.econ omist.com/business/2020/08/20/forced-labour-in-china-presents-dilemmas-for -fashion-brands.

8 **Each loaf can weigh:** Educational Resources, Harvesting, Cotton.org, https:// www.cotton.org/pubs/cottoncounts/fieldtofabric/harvest.cfm.

9 **pressed for cottonseed oil:** Cottonseed oil is $106.75 per 2,000 pounds. See *Cottonseed Intelligence Monthly* 24, issue 3 (March 4, 2020), HIS Markit, https://www .cottoninc.com/wp-content/uploads/2020/03/CIM-03-2020.pdf. Cottonseed oil is a common component of vegetable oil, which is about a tenth of the cost of olive oil. See Jody Gatewood, "Vegetable Oils–Comparison, Cost, and Nutrition," Iowa State University Extension and Outreach, August 19, 2013, https://blogs.extension

.iastate.edu/spendsmart/2013/08/19/vegetable-oils-comparison-cost-and
-nutrition/.

9 **used to be a mill in Littlefield:** KCDB Digital, "Select Milk Producers to Convert
Littlefield Denim Mill into Milk Processing Plant," October 28, 2015, https://www
.kcbd.com/story/30366617/select-milk-producers-to-convert-littlefield-denim
-mill-into-milk-processing-plant/.

12 **grow organic cotton:** "Get the Facts About Organic Cotton," https://ota.com/ad
vocacy/fiber-and-textiles/get-facts-about-organic-cotton; Sustainability Learning
Center, "Organic Cotton 101," Organic Cotton Plus, https://organiccottonplus.com
/pages/learning-center#questions-and-answers.

12 **increased by 12 percent:** "U.S. Organic Sales Break Through $50 Billion Mark in
2018," Organic Trade Association press release, May 17, 2019, https://ota.com/news
/press-releases/20699.

13 **quantity of our food:** Former war factories were also the birthplace of processed
foods, where new chemical additives, dyes, flavors, and colors created homogenous,
longer-lasting, and less expensive versions of foods people used to make themselves.
Think Wonder Bread, homogenized milk, or Spam.

15 **detrimental effects of DDT exposed:** "The Silent Spring," The Life and Legacy
of Rachel Carson, RachelCarson.org, http://www.rachelcarson.org/SilentSpring
.aspx.

15 **one hazardous pesticide:** "Is Cotton Conquering Its Chemical Addiction," A Review of Pesticide Use in Global Cotton Production, Pesticide Action Network,
October 2017, https://issuu.com/pan-uk/docs/cottons_chemical_addiction_-_final
_?e=28041656/54138689.

15 **one third of a pound of chemicals:** "Clearing Up Your Choices About Cotton,"
Natural Resources Defense Council, August 2011, https://www.nrdc.org/sites/de
fault/files/CBD_FiberFacts_Cotton.pdf.

15 **thanks to industrial farming:** R. S. Blackburn, ed., *Sustainable Textiles: Life Cycle
and Environmental Impact* (Boca Raton, FL: CRC Press, 2009), 34, https://www
.google.com/books/edition/Sustainable_Textiles/Ik6kAgAAQBAJ?hl=en&gbpv
=1&bsq=tripled.

16 **sprayed with the herbicide Roundup:** "Herbicides," CADDIS, vol. 2, United
States Environmental Protection Agency, https://www.epa.gov/caddis-vol2/caddis
-volume-2-sources-stressors-responses-herbicides.

17 **nineteen different metrics to measure soil health:** "National Soil Health Measurements to Accelerate Agricultural Transformation," Soil Health Institute, https://
soilhealthinstitute.org/national-soil-health-measurements-accelerate-agricultural
-transformation/.

18 **lower carbon footprint per unit:** Verena Seufert and Navin Ramankutty, "Many
Shades of Gray: The Context-Dependent Performance of Organic Agriculture,"
Science Advances 3, no. 3 (March 10, 2017), https://advances.sciencemag.org/content
/3/3/e1602638.

18 **advocates for regenerative agriculture:** Regenerative agriculture practices can include: no-till agriculture, where farmers avoid plowing soils and instead drill seeds
into the soil; the use of cover crops, which are plants grown to cover the soil after

farmers harvest the main crop; diverse crop rotations, such as planting three or more crops in rotation over several years; and rotating crops with livestock grazing.

18 **cut herbicide use by 90 percent:** Margy Eckelkamp, "Blue River Expands See & Spray Testing Before Commercial Launch," *Farm Journal's AGPRO*, March 5, 2018, https://www.agprofessional.com/article/blue-river-expands-see-spray-testing -commercial-launch.

18 **built up a tolerance:** "Herbicide Tolerance," University of California, Division of Agriculture and Natural Resources, UC Davis Seed Biotechnology Center, http:// sbc.ucdavis.edu/Biotech_for_Sustain_pages/Herbicide_Tolerance/.

19 **"We are essentially destroying":** Richard Schiffman, "Why It's Time to Stop Punishing Our Soils with Fertilizers," Yale Environment 360, May 3, 2017, https://e360 .yale.edu/features/why-its-time-to-stop-punishing-our-soils-with-fertilizers-and -chemicals.

19 **lax inspections:** "European Clothing Chains Hit by 'Fake' Organic Label Controversy," DW.com, https://www.dw.com/en/european-clothing-chains-hit-by-fake -organic-label-controversy/a-5164495.

20 **Some proponents of regenerative agriculture:** Regeneration International, "Regenerative Organic Agriculture and Climate Change: Down to Earth Solution to Global Warming," October 6, 2015, https://regenerationinternational.org/2015 /10/06/regenerative-organic-agriculture-and-climate-change-a-down-to-earth -solution-to-global-warming/.

20 **what keeps soil carbon sequestered:** "Creating a Sustainable Food Future," World Resources Report, https://wrr-food.wri.org/.

20 **"We need more independent research":** Schiffman, "Why It's Time to Stop Punishing Our Soils with Fertilizers."

21 **transfer ownership:** News Editor, "70 Percent of Farmland to Change Hands in Next 20 Years," American AG Radio Network, October 31, 2018, https://americanag network.com/2018/10/70-percent-of-farmland-to-change-hands-in-next-20-years/.

CHAPTER 2: TEXTILES MADE IN CHINA: HOW THE DRIVE FOR CHEAP IS KILLING THE PLANET

24 **China produced 45.86 billion meters:** "Textile Production in China from August 2019 to August 2020," Statista, Consumer Goods & FMCG, Clothing and Apparel, https://www.statista.com/statistics/226193/clothing-production-in-china-by-month/.

24 **China exported $284 billion worth of textiles:** "Statistical Tables," 2016, World Trade Organization, https://www.wto.org/english/res_e/statis_e/wts2016_e/wts16 _chap9_e.htm.

24 **in 2018, it still exported $119 billion worth of textiles:** "Statistical Tables," 2016, Table A2, World Trade Organization, https://www.wto.org/english/res_e/statis_e /wts2019_e/wts2019chapter08_e.pdf.

24 **"the world's factory":** "China Is the World's Factory, More Than Ever," *The Economist*, June 23, 2020, https://www.economist.com/finance-and-economics/2020 /06/23/china-is-the-worlds-factory-more-than-ever.

25 **more than 98 percent of clothing:** "ApparelStats 2014 and ShoeStats 2014 Reports," We Wear New, American Apparel & Footwear Association, January 9, 2015, https://web.archive.org/web/20160322062732/https://www.wewear.org/apparel stats-2014-and-shoestats-2014-reports/.

25 **1974 Multifiber Arrangement:** "Textiles Monitoring Body (TMB) The Agreement on Textiles and Clothing," World Trade Organization, https://www.wto.org /english/tratop_e/texti_e/texintro_e.htm#MFA.

25 **None of the company's jeans:** "Levi's Set to Close Last U.S. Factory," *Baltimore Sun*, October 19, 2003, https://www.baltimoresun.com/news/bs-xpm-2003-10-19 -0310190003-story.html.

25 **"Made in USA":** Meena Thiruvengadam, "Apparel Industry No Longer a Good Fit in El Paso," Institute for Agriculture & Trade Policy, October 15, 2005, https:// www.iatp.org/news/apparel-industry-no-longer-a-good-fit-in-el-paso. Until late 2017, Levi's still used a single mill in North Carolina, Cone Denim, for its "Made in the USA 501" style. Jeans still sold under this label must be remainders of that inventory. See Matt Jancer, "The Death of Denim: What the Closing of Cone Mills Means for 'Made in the USA,'" *Men's Journal*, https://www.mensjournal.com/fea tures/what-closing-cone-mills-means-made-in-the-usa/.

27 **"There is one and only one":** Milton Friedman, "The Social Responsibility of Business Is to Increase Its Profits," *New York Times*, September 13, 1970, https:// www.nytimes.com/1970/09/13/archives/article-15-no-title.html.

28 **In El Paso, a pair:** Thiruvengadam, "Apparel Industry No Longer a Good Fit in El Paso."

28 **"cut costs, drive cash flow":** Chip Bergh, "The CEO of Levi Strauss on Leading an Iconic Brand Back to Growth," *Harvard Business Review*, July–August 2018, https://hbr.org/2018/07/the-ceo-of-levi-strauss-on-leading-an-iconic-brand -back-to-growth.

28 **Philip Marineau's pay package:** WWD Staff, "Marineau's Millions: Levi's Chief Exec Sees Pay Jump to $25.1M," WWD, February 27, 2003, https://wwd.com /fashion-news/fashion-features/marineau-8217-s-millions-levi-8217-s-chief-exec -sees-pay-jump-to-25-1m-738461/.

29 **one of its Six Priorities:** Under the "Six Priorities" policy, the industry enjoyed favorable treatments in six areas: supply of raw materials, fuel, and power; innovation and its transformation and infrastructure construction; bank loans; foreign exchange; imported foreign advanced technology; and transportation.

29 **garment output in China:** Larry D. Qiu, "China's Textile and Clothing Industry" (2007).

30 **48 percent of global production:** M. R. Subramani, "How China Built $150 Billion Lead over India in Textile Exports," CPG-RMG Study 2016, March 2, 2018, http://rmg-study.cpd.org.bd/how-china-built-150-billion-lead-over -india-in-textile-exports/.

30 **300 million pairs of jeans:** "The Dirty Secret Behind Jeans and Bras," Greenpeace, December 2010, https://web.archive.org/web/20110312074819/http://www.green peace.org/eastasia/news/textile-pollution-xintang-gurao/.

30 **60 percent of all jeans:** "The Price of Success: China Blighted by Industrial Pollution—in Pictures," February 9, 2011.

30 **including Levi's:** https://www.levistrauss.com/wp-content/uploads/2019/03/Levi
-Strauss-Co-Factory-Mill-List-March-2019.pdf.

30 **and Abercrombie:** https://corporate.abercrombie.com/af-cares/sustainability/social
/audit-lifecycle/active-factory-list.

35 **In China, 85.7 percent of the energy grid:** *China Statistical Yearbook*, 2018 data,
Energy, Table 9-2: Total Consumption of Energy and Its Composition, http://
www.stats.gov.cn/tjsj/ndsj/2019/indexeh.htm.

35 **63 percent:** "What Is U.S. Electricity Generation by Energy Source," U.S. Energy
Information Administration, last updated February 27, 2020, https://www.eia
.gov/tools/faqs/faq.php?id=427&t=3.

35 **78 tons of CO_2 per unit:** "Brown to Green: The G20 Transition Towards a Net-
Zero Emissions Economy, 2019: China," https://www.climate-transparency.org/wp
-content/uploads/2019/11/B2G_2019_China.pdf.

35 **compared with the United States' 60:** "Brown to Green: The G20 Transition
Towards a Net-Zero Emissions Economy, 2019: Mitigation Energy: United States,"
https://www.climate-transparency.org/wp-content/uploads/2019/11/B2G_2019
_USA.pdf#page=4.

35 **the European Union's 54 tons:** "Brown to Green: The G20 Transition Towards a
Net-Zero Emissions Economy, 2019: Mitigation Energy: European Union," https://
www.climate-transparency.org/wp-content/uploads/2019/11/B2G_2019_EU
.pdf#page=4.

35 **clothing industry's carbon footprint:** "Measuring Fashion: Environmental Impact
of the Global Apparel and Footwear Industries Study," Quantis, 2018, https://
quantis-intl.com/wp-content/uploads/2018/03/measuringfashion_globalimpact
study_full-report_quantis_cwf_2018a, for the over 75 percent figure and for the
8.1 figure, and Achim Berg and Karl-Hendrik Magnu, "Fashion on Climate,"
McKinsey & Company, 2020, https://www.mckinsey.com/-/media/McKinsey
/Industries/Retail/Our%20Insights/Fashion%20on%20climate/Fashion-on
-climate-Full-report, for the 4 percent figure. It is troubling that neither of these
reports disclose their underlying data. That these reports have such wildly different
results is a demonstration that more quality and transparent information is needed
to accurately assess the impact of this industry.

35 **fashion contributes the same level of greenhouse gases:** Berg and Magnu, "Fashion on Climate."

35 **clothes would use 26 percent:** "A New Textiles Economy: Redesigning Fashion's
Future," Ellen MacArthur Foundation, https://www.ellenmacarthurfoundation
.org/assets/downloads/publications/A-New-Textiles-Economy_Summary-of
-Findings_Updated_1-12-17.pdf.

36 **US apparel imports from China:** Dr. Sheng Lu, "COVID-19 and U.S. Apparel
Imports," FASH455 Global Apparel & Textile Trade and Sourcing, updated
September 2020, https://shenglufashion.com/2020/09/04/covid-19-and-u-s-apparel
-imports-updated-september-2020/.

36 **Guangdong is one of the three largest:** IBP, Inc., *China: Clothing and Textile Industry Handbook: Strategic Information and Contacts*, 2016 edition, https://books
.google.com/books?id=CgWtDwAAQBAJ&pg=PA127&lpg=PA127&dq=Guang
dong+textile+volume&source=bl&ots=7tdVmWwNpz&sig=ACfU3U2b_Nf8Q

SoIxlT8Wa-TyzCvr_ZVbw&hl=en&sa=X&ved=2ahUKEwjdqKrIv8nkAhUUtX
EKHUG3BjgQ6AEwA3oECAgQAQ#v=onepage&q=volume&f=false.

36 **It had sixty thousand factories:** Hsiao-Hung Pai, "Factory of the World: Scenes from Guangdong," *Places Journal*, October 2012, https://placesjournal.org/article/factory-of-the-world-scenes-from-guangdong/?cn-reloaded=1. An unknown number of factories have closed, and reopened following a crackdown on CO_2 emissions.

38 **classifies it as a "danger":** "Potassium Permanganate," IPCS Inchem, http://www.inchem.org/documents/icsc/icsc/eics0672.htm.

39 **China Water Risk estimate:** Debra Tan, "Dirty Thirsty Wars—Fashion Blindsided," CWR, September 17, 2014, https://www.chinawaterrisk.org/resources/analysis-reviews/dirty-thirsty-wars-fashion-blindsided/.

41 **While there is evidence:** "How Is China Managing Its Greenhouse Gas Emissions?," ChinaPower, https://chinapower.csis.org/china-greenhouse-gas-emissions/.

41 **1.2 million premature deaths from air pollution:** "State of Global Air/2019," Health Effects Institute 2019, State of Global Air 2019, Special Report (Boston: Health Effects Institute), https://www.stateofglobalair.org/sites/default/files/soga_2019_report.pdf.

42 **factories getting shut down:** "State of Global Air/2019," 6.

43 **the Clean Water Act:** It has to be noted that the Trump administration has worked to dismantle a lot of this, though not to the extent of reverting back to 1960s levels.

43 **We just shipped them:** Jason Hickel, "The Myth of America's Green Growth," *Foreign Policy*, June 18, 2020, https://foreignpolicy.com/2020/06/18/more-from-less-green-growth-environment-gdp/.

44 **The rivers I visited:** Tom Phillips, "A 'Black and Smelly' Job: The Search for China's Most Polluted Rivers," *Guardian*, June 21, 2016, https://www.theguardian.com/world/2016/jun/22/black-smelly-citizens-clean-chinas-polluted-rivers.

44 **initiatives by Greenpeace:** "A Monstrous Mess: Toxic Water Pollution in China," Greenpeace, January 23, 2014, https://www.greenpeace.org/international/story/6846/a-monstrous-mess-toxic-water-pollution-in-china/.

45 **ten hot-water baths:** Ayşe Merve Kocabaş, "Improvements in Energy and Water Consumption Performances of a Textile Mill After Bat Applications," https://etd.lib.metu.edu.tr/upload/12609296/index.pdf.; Maria Laura Parisi et al., "Environmental Impact Assessment of an Eco-Efficient Production for Coloured Textiles," *Journal of Cleaner Production* 108, part A, 514–24, https://www.sciencedirect.com/science/article/abs/pii/S095965261500757X?via%3Dihub.

45 **German Catalogue of Textile Auxiliaries:** "Introduction to the Problems Surrounding Garment Textiles," updated BfR Opinion No. 041/2012, *Bundesinstitut für Risikobewertung*, July 6, 2012, https://www.bfr.bund.de/cm/349/introduction-to-the-problems-surrounding-garment-textiles.pdf.

45 **28 percent chemicals by weight:** The literature does suggest that there are more toxic chemicals found in synthetics versus natural fibers, but it's not black and white. Giovanna Luongo, "Chemicals in Textiles: A Potential Source for Human Exposure and Environmental Pollution," PhD diss., Stockholm University, 2015.

45 **two thirds of the world's population:** "Water Scarcity," International Decade for Action "Water for Life" 2005–2015, United Nations Department of Economic and

Social Affairs, last updated November 24, 2014, https://www.un.org/waterforlifede
cade/scarcity.shtml.

45 **printing and dyeing wastewater:** Yi Li et al., "Water Environmental Stress, Re-
bound Effect, and Economic Growth of China's Textile Industry," *PeerJ* (June 29,
2018), https://www.ncbi.nlm.nih.gov/pmc/articles/PMC6056267/.

45 **water for humans, livestock:** Adam Matthews, "The Environmental Crisis in Your
Closet," *Newsweek*, August 13, 2015, https://www.newsweek.com/2015/08/21/envi
ronmental-crisis-your-closet-362409.html.

46 **the harm that effluents cause:** Rita Kant, "Textile Dyeing Industry an Environmental Hazard," *Natural Science* 4, no. 1 (2012), https://file.scirp.org/Html/4-8301582
_17027.htm.

46 **approximately 10 percent:** Go4more.global, Dr. Reiner Hengstmann, and Char-
maine Nuguid, "Input Paper on Private Standards, Labels and Certification
Mechanisms in the Post-2020 Chemicals and Waste Framework," Third Meeting
of the SAICM Intersessional, Bangkok, Thailand, September 30 to October 4, 2019,
http://saicm.org/Portals/12/documents/meetings/IP3/INF/SAICM_IP3_INF
_11_Role_private_standards_GOV_Switzerland.pdf.

46 **Polluted water has been found:** Joaquim Rovira and José L. Domingo, "Human
Health Risk Due to Exposure to Inorganic and Organic Chemicals from Textiles:
A Review," *Environmental Research* 168 (2019): 62–69.

46 **the food chain:** Bruno Lellis et al., "Effects of Textile Dyes on Health and the
Environment and Bioremediation Potential of Living Organisms," *Biotechnology
Research and Innovation* 3, no. 2 (July–December 2019): 275–90, https://www
.sciencedirect.com/science/article/pii/S2452072119300413.

46 **common among them being starch:** "Starch Wastewater Treatment Solution," NGO
International, http://ngoenvironment.com/en/Types-of-wastewater-tec34-STARCH
-WASTEWATER-TREATMENT-SOLUTION-d133.html/.

46 **the starch depletes the oxygen:** Mahmmoud Nasr, "Biological Hydrogen Production from Starch Wastewater Using a Novel Up-flow Anaerobic Staged Reactor,"
Table 1: Characteristics of the Starch Processing Wastewater Used in the Experi-
ments, *BioResources* 8, no. 4, https://www.researchgate.net/figure/Characteristics-of
-the-Starch-Processing-Wastewater-Used-in-the-Experiments_tbl1_259810426.

47 **chemicals that rub up against our bodies:** J. Xue, W. Liu, and K. Kannan, "Bi-
sphenols, Benzophenones, and Bisphenol A Diglycidyl Ethers in Textiles and Infant Clothing," *Environmental Science & Technology* 51, no. 9 (May 2, 2017):
5279–86.

48 **illnesses like bladder cancer:** Zorawar Singh and Pooja Chadha, "Textile Industry and Occupational Cancer," *Journal of Occupational Medicine and Toxicology*
11, no. 39 (August 15, 2016), https://www.ncbi.nlm.nih.gov/pmc/articles/PMC
4986180/.

48 **once they get to us:** David Ewing Duncan, "Chemicals Within Us," *National Geo-
graphic*, n.d., https://www.nationalgeographic.com/science/health-and-human-body
/human-body/chemicals-within-us/.

49 **some of their pseudonyms:** Marc Bain, "If Your Clothes Aren't Already Made Out
of Plastic, They Will Be," Quartz.com, June 5, 2015, https://qz.com/414223/if-your
-clothes-arent-already-made-out-of-plastic-they-will-be/.

49 **Wallace H. Carothers, who discovered:** "Nylon Is Invented, 1935," PBS, A Science Odyssey: People and Discoveries, http://www.pbs.org/wgbh/aso/databank /entries/dt35ny.html.

50 **John Whinfield and James Dickson:** "What Is Polyester: History of Polyester," WhatIsPolyester.com, http://www.whatispolyester.com/history.html.

50 **it is not a global competitor:** News Desk, "Global Polyester Yarn Exports Rising Since 2017," Fibre2Fashion.com, October 5, 2019, https://www.fibre2fashion.com /news/textile-news/global-polyester-yarn-exports-rising-since-2017-252377 -newsdetails.htm.

50 **China has also come to dominate polyester fiber production:** Yan Qin, "Global Fibres Overview," Synthetic Fibres Raw Materials Committee Meeting at APIC 2014, May 16, 2014, Tecnon Orbichem, https://www.orbichem.com/userfiles/APIC %202014/APIC2014_Yang_Qin.pdf.

51 **fiber production stage:** Quantis, "Measuring Fashion: Environmental Impact of the Global Apparel and Footwear Industries Study," 2018, https://https://quantis -intl.com/report/measuring-fashion-report/.

51 **polyester has a larger footprint:** Business for Social Responsibility, "Apparel Industry Life Cycle Carbon Mapping," June 2009, BSR.org, https://www.slideshare .net/AbhishekBhagat1/bsr-apparel-supplychaincarbonreport.

51 **form of microplastics:** To clarify, microfibers are not the same as microplastics, words you've probably come across in those apocalyptic news stories. Microfibers are the fibers that shed from any type of clothing, synthetic or natural, during production, our own use, or end of life, and end up as pollution in the environment. Microplastics, by contrast, is a term for any plastic particle 5 millimeters or less, from clothing or other things, too. Synthetic microfibers are a type of microplastic, but not all microfibers are microplastics, and not all microplastics are microfibers (think: microbeads, those little scrubbies that promised to get rid of blackheads and were banned from our cosmetic products because of their impact on the environment).

51 **more plastic than fish:** Sarah Kaplan, "By 2050, There Will Be More Plastic Than Fish in the World's Oceans, Study Says," *Washington Post*, January 20, 2016, https://www.washingtonpost.com/news/morning-mix/wp/2016/01/20/by-2050 -there-will-be-more-plastic-than-fish-in-the-worlds-oceans-study-says/.

52 **the plastic crisis is largely invisible:** Yogis, take note—the leggings you don to keep your body and mind healthy, especially those made of "recycled" plastic bottles, are major culprits of microplastic shed. Just take a look at your pair before your next practice. If you see small whitish fuzz, there's your petroleum!

52 **6 percent of the total mass:** Beverley Henry et al., "Microfibres from Apparel and Home Textiles: Prospects for Including Microplastics in Environmental Sustainability Assessment," *Science of the Total Environment* 652 (February 20, 2019): 484–94, https://www.sciencedirect.com/science/article/pii/S004896971834049X CQ.

52 **synthetic microfibers have been found:** Julien Boucher and Damien Friot, "Primary Microplastics in the Oceans: a Global Evaluation of Sources," Gland, Switzerland: International Union for Conservation of Nature, 2017, 43.

52 **more than 1,900 microfibers:** Mark Anthony Browne et al., "Accumulations of Microplastic on Shorelines Worldwide: Sources and Sinks," *Environmental Science & Technology* (September 6, 2011), DOI: 10.1021/es201811s, https://www.plastic

soupfoundation.org/wp-content/uploads/2015/03/Browne_2011-EST
-Accumulation_of_microplastics-worldwide-sources-sinks.pdf.

52 **more than 700,000 released:** Imogene E. Napper and Richard C. Thompson, "Release of Synthetic Microplastic Plastic Fibres from Domestic Washing Machines: Effects of Fabric Type and Washing Conditions," *Marine Pollution Bulletin* 112, no. 1–2 (November 15, 2016): 39–45, http://www.inquirylearningcenter.org /wp-content/uploads/2015/08/Napper2016.pdf.

52 **209,000 tons of synthetic microfibers:** Beverley Henry, Kirsi Laitala, and Ingun Grimstad Klepp, "Microfibres from Apparel and Home Textiles: Prospects for Including Microplastics in Environmental Sustainability Assessment," *Science of the Total Environment* 652 (2019): 483–94, https://www.sciencedirect.com/science /article/pii/S004896971834049X.

52 **found to reproduce less:** Andrea Thompson, "From Fish to Humans, a Microplastic Invasion May Be Taking a Toll," *Scientific American*, September 4, 2018, scientific american.com/article/from-fish-to-humans-a-microplastic-invasion-may-be -taking-a-toll/.

CHAPTER 3: MY FACTORY IS A CAGE: CUT AND SEW AND THE CRISIS OF LABOR

56 **62 percent of chief purchasing officers:** "The Apparel Sourcing Caravan's Next Stop: Digitization," McKinsey Apparel CPO Survey 2017, Apparel, Fashion & Luxury Group, McKinsey & Company, https://www.mckinsey.com/~/media/mckinsey /industries/retail/our%20insights/digitization%20the%20next%20stop%20for %20the%20apparel%20sourcing%20caravan/the-next-stop-for-the-apparel -sourcing-caravan-digitization.pdf.

56 **China was still the leading garment exporter:** "World Trade Statistical Review 2020," World Trade Organization 2020, https://www.wto.org/english/res_e/sta tis_e/wts2020_e/wts2020_e.pdf.

56 **guess who? Bangladesh:** Jasim Uddin, "Bangladesh's Denim Overtakes Mexico, China, in US," *Business Standard*, August 11, 2020, https://tbsnews.net/economy /rmg/bangladeshs-denim-overtakes-mexico-china-us-118252.

57 **80 percent of the country:** H. Brammer, "Floods in Bangladesh: Geographical Background to the 1987 and 1988 Floods," *Geographical Journal* 156, no. 1 (March 1990): 12–22, https://www.jstor.org/stable/635431.

57 **world's most vulnerable countries:** "Bangladesh," Aquastat, FAO.org, http://www .fao.org/nr/water/aquastat/countries_regions/bgd/BGD-CP_eng.pdf.

57 **188 percent expansion in GDP:** Sheikh Hasina, "Bangladesh Is Booming—and Here's Why, Says the Prime Minister," World Economic Forum, October 4, 2019, https://www.weforum.org/agenda/2019/10/bangladesh-is-booming/.

57 **population below the poverty line:** "The World Bank in Bangladesh," World Bank, last updated October 14, 2020, https://www.worldbank.org/en/country/bangladesh /overview.

58 **average lifespan has increased:** Fiona Weber-Steinhaus, "The Rise and Rise of Bangladesh," *The Guardian*, October 9, 2019, https://www.theguardian.com/

global-development/2019/oct/09/bangladesh-women-clothes-garment-workers
-rana-plaza.

58 **Bangladeshi equivalents earned:** "Wages and Productivity in Garment Sector in Asia and the Pacific and the Arab States," International Labour Organization, n.d., https://www.ilo.org/wcmsp5/groups/public/—asia/—ro-bangkok/documents /publication/wcms_534289.pdf.

59 **$26 per month salary:** Abdi Latif Dahir, "Ethiopia's Garment Workers Make Clothes for Guess, H&M, and Levi's—but Are the World's Lowest Paid," Quartz .com, May 8, 2019, https://qz.com/africa/1614752/ethiopia-garment-workers-for -gap-hm-lowest-paid-in-world/.

59 **the industry now comprises:** Shahajida Mia and Masrufa Akter, "Ready-Made Garments Sector of Bangladesh: Its Growth, Contribution and Challenges," *Economics World* 7, no. 1 (January–February 2019): 17–26, http://www.davidpublisher .org/Public/uploads/Contribute/5dd507c82e7dd.pdf.

61 **what a black river looks like:** Tanvir Chowdhury, "Bangladesh's Garment Factories Pollute Rivers," Aljazeera, July 1, 2019, https://www.aljazeera.com/news/2019/07 /bangladeshs-garment-factories-pollute-rivers-affecting-residents-health -190701090533205.html.

61 **Three of Bangladesh's rivers:** This same process was seen in England some sixty years ago, when the River Thames was famously declared "biologically dead." It has since been revitalized and is now one of the cleanest rivers flowing through a major city. See Sophie Hardach, "How the River Thames Was Brought Back from the Dead," BBC, November 12, 2015, http://www.bbc.com/earth/story/20151111 -how-the-river-thames-was-brought-back-from-the-dead.

63 **80 percent of cut-and-sew workers:** See for example, "The Women Who Make Our Clothes," FashionRevolution.org, n.d., https://www.fashionrevolution.org/asia -vietnam-80-percent-exhibition/; and "Gender: Women Workers Mistreated," Clean Clothes Campaign, n.d., https://cleanclothes.org/issues/gender.

68 **Fires had ravaged this area:** Salman Saeed, Sugam Pokharel, and Matthew Robinson, "Bangladesh Slum Fire Leaves 10,000 People Homeless," CNN, August 19, 2019, https://edition.cnn.com/2019/08/18/asia/dhaka-bangladesh-slum-fire-10000 -homeless-intl/index.html.

70 **a compelling UNICEF fund-raising video:** "UNICEF in Bangladesh: Our Timeline," UNICEF.org, https://www.unicef.org/bangladesh/en/unicef-bangladesh.

71 **sixty-two hours per week:** "The Bangladeshi Garment Worker Diaries," Worker Diaries.org, February 13, 2018, https://workerdiaries.org/wp-content/uploads/2018 /04/Bangladesh_Data_Portal_English.pdf.

71 **64 percent of garment workers:** "The Bangladeshi Garment Worker Diaries," 21.

71 **government-set minimum wage:** "Living Wage in Asia," Clean Clothes Campaign, 2014, https://archive.cleanclothes.org/resources/publications/asia-wage-report/view.

71 **77 percent were anemic:** Taslima Khatun et al., "Anemia Among Garment Factory Workers in Bangladesh," *Middle East Journal of Scientific Research* 16, no. 4 (January 2013): 502–7, https://www.researchgate.net/publication/263027531 _Anemia_among_Garment_Factory_Workers_in_Bangladesh. "Research carried out by UK Clean Clothes Campaign, Labour Behind the Label, has found that

factory workers consume an average of 1598 calories a day, around half the recommended amount. Body Mass Index figures indicated 33% of Cambodian workers are medically underweight and at risk, and 25% seriously so, displaying figures that would be used to diagnose anorexia in the UK." Clean Clothes Campaign, "Living Wage in Asia" (2014), 39.

72 **67 percent reported they were physically assaulted:** Nusrat Zaman Sohani et al., "Pattern of Workplace Violence Against Female Garment Workers in Selected Areas of Dhaka City," *Sub Journal of Public Health* 3–4, 3–8 (July–December 2010/January–June 2011), https://www.researchgate.net/publication/233883183 _PATTERN_OF_WORKPLACE_VIOLENCE_AGAINST_FEMALE_GAR MENT_WORKERS_IN_SELECTED_AREAS_OF_DHAKA_CITY.

72 **nearly eight thousand were dismissed:** "Bangladesh: Investigate Dismissals of Protesting Workers," Human Rights Watch, March 5, 2019, https://www.hrw.org /news/2019/03/05/bangladesh-investigate-dismissals-protesting-workers.

73 **a bunch of papers and memoirs:** Barbara Ehrenreich, *Nickel and Dimed: On (Not) Getting By in America* (New York: Henry Holt and Co., 2001) is an excellent examination of life for laborers.

73 **"There is no real question":** Sven Beckert, *Empire of Cotton: A Global History* (New York: Knopf, 2014), 16.

74 **women were forced into sex trafficking:** Sex trafficking was reported in cases of garment workers attempting to travel to a factory, or to secure a job as a garment worker. See K. Amirthalingam et al., "Victims of Human Trafficking in Sri Lanka: Narratives of Women, Children and Youth," EditorialExpress.com, https://editorialexpress.com /cgi-bin/conference/download.cgi?db_name=IAFFE2011&paper_id=75.

74 **garment industry in Cambodia:** US Mission Cambodia, "2020 Trafficking in Persons Report: Cambodia," US Embassy in Cambodia, News and Events, July 16, 2020, https://kh.usembassy.gov/2020-trafficking-in-persons-report-cambodia / https://www.voacambodia.com/a/author-looks-at-forced-labor-in-cambodias-sex -and-garment-industries/2726966.html.

75 **"rescued" from the sex trade:** Patrick Winn, "Why Cambodia's Sex Workers Don't Need to Be Saved," *GlobalPost*, March 29, 2016, PRI.org, https://www.pri .org/stories/2016-03-29/why-cambodias-sex-workers-dont-need-be-saved.

76 **earn only between .5 percent and 4 percent:** "There Is a Vast Mismatch," Behind the Seams, http://behind-the-seams.org/living-wages/.

76 **If H&M raised the cost:** "What She Makes: Power and Poverty in the Fashion Industry," OXFAM Australia, October 2017, https://whatshemakes.oxfam.org.au /wp-content/uploads/2017/10/Living-Wage-Media-Report_WEB.pdf.

76 **worth more than $17 billion:** #81 Stefan Persson, Billionaires 2020, *Forbes*, November 14, 2020, https://www.forbes.com/profile/stefan-persson/#27fae5315dbe.

76 **net worth of around $64.9 billion:** #6 Amancio Ortega, https://www.forbes.com /profile/amancio-ortega/#46c51314116c.

76 **briefly beat out Bill Gates:** Kate Vinton, "Briefly No. 1: Spain's Amancio Ortega Ends Day Back at World's No. 2 Richest," *Forbes*, August 29, 2017, https://www .forbes.com/sites/katevinton/2017/08/29/spains-amancio-ortega-briefly -overtakes-gates-as-no-1-richest-falls-back-to-no-2/#54f80cd667be.

78 **US accepts more than $400 billion:** Jason Fields, "US Ban on Slave-Made Goods Nets Tiny Fraction of $400 Billion Threat," Thomson Reuters Foundation, April 8, 2019, https://news.trust.org/item/20190408044809-3ud9e/.

79 **second-highest amount of money:** "Modern Slavery: A Hidden, Everyday Problem," Global Slavery Index, 2018, https://www.globalslaveryindex.org/.

79 **the United States seized only $6.3 million:** Kieran Gilbert, "U.S. Blocks Import of Goods from Five Nations in Rare Anti-Slavery Crackdown," Reuters, October 1, 2019, https://news.yahoo.com/1-u-blocks-import-goods-122024862.html.

79 **US Customs and Border Protection:** Fields, "US Ban on Slave-Made Goods."

79 **that team consists of just six employees:** Fields, "US Ban on Slave-Made Goods."

79 **Europe offers an example:** "French Corporate Duty of Vigilance Law," European Coalition of Corporate Justice, 2016, Respect International, http://www.respect .international/french-corporate-duty-of-vigilance-law-english-translation/.

79 **the law could fine companies:** Jane Moyo, "France Adopts New Corporate 'Duty of Care' Law, Ethical Trading Initiative," EthicalTrade.org, March 1, 2017, https:// www.ethicaltrade.org/blog/france-adopts-new-corporate-duty-care-law.

79 **responsible for paying compensation:** "France Duty of Vigilance Law," WorkerEngagement.com, https://www.worker-engagement.com/laws-regulations-and-guidelines /french-duty-of-vigilance-duty-of-care-law/.

81 **help make New York the first American epicenter:** Ben Passikoff, *The Writing on the Wall: Rediscovering New York City's "Ghost Signs"* (New York: Simon & Schuster, 2017).

81 **produced in New York:** Jacob Riis, *How the Other Half Lives,* "Knee-Pants at Forty-Five Cents a Dozen—A Ludlow Street Sweater's Shop," https://www.khanacademy .org/humanities/art-americas/us-art-19c/us-19c-arch-sculp-photo/a/jacob-riis -sweaters.

81 **Jews from Eastern Europe:** Yannay Spitzer, "Pogroms, Networks, and Migration: The Jewish Migration from the Russian Empire to the United States, 1881–1914," *Brown University* 29 (2015).

82 **successfully procure legislation:** "Ten Hours Act," Oxford Reference, https:// www.oxfordreference.com/view/10.1093/oi/authority.20110803103058890.

82 **biggest women's strike:** Tony Michels, "Uprising of 20,000 (1909)," Jewish Women's Archive, https://jwa.org/encyclopedia/article/uprising-of-20000-1909.

82 **unified banner of feminism:** Eileen Boris and Annelise Orleck, "Feminism and the Labor Movement: A Century of Collaboration and Conflict," *New Labor Forum* 20, no. 1 (Winter 2011): 33–41, https://www.jstor.org/stable/27920539.

82 **Rose Schneiderman, a Jewish immigrant:** Susan Ware, "A Strong Working-Class Movement: On the Activism of Rose Schneiderman," *Lapham's Quarterly,* May 9, 2019, https://www.laphamsquarterly.org/roundtable/strong-working-class-movement.

83 **Triangle Shirtwaist Factory:** Patrick J. Kiger, "How the Horrific Tragedy of the Triangle Shirtwaist Fire Led to Workplace Safety Laws," History.com, updated March 27, 2019, https://www.history.com/news/triangle-shirtwaist-factory-fire-labor -safety-laws.

83 **sixty-two girls jumped:** Clair T. Berube, *The Investments: An American Conspiracy* (Charlotte, NC: Information Age Publishing, 2020), 6.

83 **One hundred and forty-six:** Ware, "A Strong Working-Class Movement."

83 **The commission resulted in legislation:** "Report of the New York State Factory Investigating Commission," *Monthly Review of the U.S. Bureau of Labor Statistics* 2, no. 2 (February 1916): 81–99, https://www.jstor.org/stable/41822920?seq=1#metadata_info_tab_contents.

83 **the New Deal began with:** Christopher N. Breiseth, "From the Triangle Fire to the New Deal: Frances Perkins in Action," speech in commemoration of the 100th anniversary of the Triangle Fire, New York State Museum, Albany, New York, March 25, 2011, https://francesperkinscenter.org/wp-content/uploads/2014/04/from-the-triangle-fire-to-the-new-deal.pdf.

88 **Farah employed 9,500 workers:** Myrna Zanetell, "Farah, Incorporated," Texas State Historical Association Handbook of Texas, https://www.tshaonline.org/handbook/entries/farah-incorporated.

88 **El Paso lost 22,000:** Meena Thiruvengadam, "Apparel Industry No Longer a Good Fit in El Paso," Institute for Agriculture & Trade Policy, October 15, 2005, https://www.iatp.org/news/apparel-industry-no-longer-a-good-fit-in-el-paso.

88 **5 million manufacturing jobs:** Patricia Atkins et al., "Responding to Manufacturing Job Loss: What Can Economic Development Policy Do?," Brookings Institution, Metropolitan Policy Program, June 2011, https://www.brookings.edu/wp-content/uploads/2016/06/06_manufacturing_job_loss.pdf.

88 **included skills-based trainings:** Andrew Stettner, "Should Workers Facing Technological Change Have a Right to Training?," The Century Foundation, September 12, 2019, https://tcf.org/content/commentary/workers-facing-technological-change-right-training/.

89 **NAFTA could have included protections:** Anand Giridharadas, *Winners Take All: The Elite Charade of Changing the World* (New York: Alfred A. Knopf, 2018), 238.

89 **greater exposure to losses from globalization:** David Autor et al., "Importing Political Polarization? The Electoral Consequences of Rising Trade Exposure," Massachusetts Institute of Technology, February 2020, https://economics.mit.edu/files/11559.

89 **$15 minimum wage by 2022:** "Minimum Wage," State of California Department of Industrial Relations, Labor Commissioner's Office, https://www.dir.ca.gov/dlse/faq_minimumwage.htm.

90 **$15 minimum wage in July 2020:** "Los Angeles Minimum Wage," Office of Wage Standards, https://wagesla.lacity.org/sites/g/files/wph471/f/2019-MWO-Poster-EN-14.pdf https://wagesla.lacity.org.

90 **all-time low of 6.2 percent:** "Union Members Summary," Economic News Release, U.S. Bureau of Labor Statistics, 2019, https://www.bls.gov/news.release/union2.nr0.htm.

90 **when organizing was illegal before 1935:** Gerald Mayer, "Union Membership Trends in the United States, Washington, D.C.: Congressional Research Service, August 21, 2004, https://digitalcommons.ilr.cornell.edu/cgi/viewcontent.cgi?article=1176&context=key_workplace.

90 **We are far behind:** Matthew Desmond, "In Order to Understand the Brutality of American Capitalism, You Have to Start on the Plantation," *New York Times*, August 14, 2019, https://www.nytimes.com/interactive/2019/08/14/magazine/slavery-capitalism.html.

91 **In Germany, for instance:** Justin Fox, "Why German Corporate Boards Include Workers," *Bloomberg*, August 24, 2018, https://www.bloomberg.com/opinion/articles/2018-08-24/why-german-corporate-boards-include-workers-for-co-determination.

91 **unions are often company-based:** Andrew Brooks, *Clothing Poverty: The Hidden World of Poverty and Second-hand Clothes* (London: Zed Books, 2015), 246.

91 **The United States was last:** "Strictness of Employment Protection—Individual and Collective Dismissals (Regular Contracts)," OECD Stats, https://stats.oecd.org/Index.aspx?DataSetCode=EPL_OV.

92 **In a hyperglobalized world:** Marina N. Bolotnikova, "The Trilemma," *Harvard Magazine*, July–August 2019, https://www.harvardmagazine.com/2019/07/rodrik-trilemma-trade-globalization.

92 **Corruption Perceptions Index:** Corruption Perception Index 2018, Transparency International, Berlin, 2018, https://images.transparencycdn.org/images/CPI_2018_Executive_Summary_EN.pdf.

CHAPTER 4: MIDDLEMEN, MANAGEMENT, MARKETING, AND A NEW KIND OF TRANSPARENCY

94 **says "Made in China":** If a product bears the label "Made in USA" it is supposed to mean that both the materials (fabric) and the production (cut and sew) happens in the United States. For products made outside of North America the rule is different. In those instances, the label only indicates the last country in which the product was substantially transformed, which, in the case of clothing, would be at the cut-and-sew stage. See "Says 'Made in China,' U.S. Rules of Origin," U.S. Customs and Border Protection, May 2004, https://www.cbp.gov/sites/default/files/assets/documents/2016-Apr/icp026_3.pdf.

94 **Li & Fung is the largest:** Nikki Sun, "Hong Kong's Li & Fung Taps JD.com for Digital Supply Chain," *Nikkei Asia*, July 31, 2020, https://asia.nikkei.com/Business/Business-deals/Hong-Kong-s-Li-Fung-taps-JD.com-for-digital-supply-chain-revamp.

95 **Founded in 1906:** "Our History," Li & Fung, lifung.com, https://www.lifung.com/about-lf/our-purpose/our-history/.

95 **sales surpassed $12 billion:** Robert J. S. Ross et al., "A Critical Corporate Profile of Li & Fung," Mosakowski Institute for Public Enterprise, 31, https://commons.clarku.edu/cgi/viewcontent.cgi?article=1030&context=mosakowskiinstitute.

95 **"We might decide to buy yarn":** Joan Magretta, "Fast, Global, and Entrepreneurial: Supply Chain Management, Hong Kong Style," *Harvard Business Review*, September–October 1998, https://hbr.org/1998/09/fast-global-and-entrepreneurial-supply-chain-management-hong-kong-style.

95 **"optimizing each step":** Magretta, "Fast, Global, and Entrepreneurial."

95 **companies such as Walmart:** Ross et al., "A Critical Corporate Profile of Li & Fung."

96 **Kathie Lee Gifford scandal:** Stephanie Strom, "A Sweetheart Becomes Suspect: Looking Behind Those Kathie Lee Labels," *New York Times*, June 27, 1996, https://

www.nytimes.com/1996/06/27/business/a-sweetheart-becomes-suspect-looking
-behind-those-kathie-lee-labels.html.

96 **Nike also had a PR nightmare:** Jennifer Burns, "Hitting the Wall: Nike and International Labor Practices," *Harvard Business Review*, January 19, 2000, https://store.hbr.org/product/hitting-the-wall-nike-and-international-labor-practices/700047?sku=700047-PDF-ENG.

96 **"these codes have become":** "We Go as Far as Brands Want Us to Go," Clean Clothes Campaign, n.d., https://cleanclothes.org/news/2019/we-go-as-far-as-brands-want-us-to-go.

97 **"At Everlane, we want":** "We Believe We Can All Make a Difference," Everlane.com, https://www.everlane.com/about.

97 **Merchant brand stalwart Gap:** "Good Business Can Change the World," Gap Inc. Global Sustainability, https://www.gapincsustainability.com/.

97 **"At Madewell, we strive":** "Social Responsibility," Madewell.com, https://www.madewell.com/social-responsibility.html.

97 **"We imagine a world":** "Our Social Responsibility," J.Crew.com, https://www.jcrew.com/flatpages/social_responsibility2019.jsp.

97 **J.Crew Group filed for bankruptcy:** Vanessa Friedman, Sapna Maheshwar, and Michael J. de la Merced, "J. Crew Files for Bankruptcy in Virus's First Big Retail Casualty," *New York Times,* May 3, 2020, https://www.nytimes.com/2020/05/03/business/j-crew-bankruptcy-coronavirus.html/.

97 **"People are at the heart":** "How We Do Business," Inditex.com, https://www.inditex.com/en/our-commitment-to-people.

97 **"aims to conduct business":** "Vendor Code of Conduct," Everlane.com, https://www.everlane.com/vendor-code.

98 **worth around $80 billion:** https://views-voices.oxfam.org.uk/2011/08/buyers-beware-audit-idiocy.

99 **"ethical sourcing" budgets:** Juliane Reineck et al., "Business Models and Labour Standards: Making the Connection," Ethical Trade Initiative, 14, https://www.ethicaltrade.org/sites/default/files/shared_resources/Business models %26 labour standards.pdf.

99 **50 percent increase:** "Fig Leaf for Fashion: How Social Auditing Protects Brands and Fails Workers," 2019 Report, Clean Clothes Campaign, 52.

99 **a factory in Bangladesh:** Siobhan Heanue, "Lululemon Factory Workers Allegedly Subjected to Physical and Verbal Abuse in Bangladesh," ABC.net.au, October 15, 2019, https://www.abc.net.au/news/2019-10-15/lululemon-abuse-allegations-women-bangladesh-factories/11605468.

99 **"We require that all vendors":** Sarah Marsh and Redwan Ahmed, "Workers Making £88 Lululemon Leggings Claim They Are Beaten," *Guardian*, October 14, 2019, https://www.theguardian.com/global-development/2019/oct/14/workers-making-lululemon-leggings-claim-they-are-beaten.

99 **When the Brazilian government:** Stephen Burgen and Tom Phillips, "Zara Accused in Brazil Sweatshop Inquiry," *Guardian*, August 18, 2011, https://www.theguardian.com/world/2011/aug/18/zara-brazil-sweatshop-accusation.

100 **sixty hours a week:** Jonathan Webb, "Child Workers Found in Clothing Supply Chain: ASOS, Marks & Spencer Implicated," *Forbes,* October 25, 2016, https://

www.forbes.com/sites/jwebb/2016/10/25/child-workers-found-in-clothing
-supply-chain-asos-marks-spencer-implicated/#614e19424b12.

100 **audits are arduous:** Jasmin Malik Chua, "Why Tackling 'Audit Fatigue' Can Lead
to More Sustainable Factories," *Sourcing Journal*, September 9, 2019, https://sourcing
journal.com/topics/sustainability/audit-fatigue-factories-sustainability-166443/.

100 **work to streamline this:** "Converged Assessment. Collaborative Action. Improved
Working Conditions," Social & Labor Convergence, https://slconvergence.org/.

100 **twelve factories used by Gap:** "Fig Leaf for Fashion."

102 **over 60 percent of suppliers:** *Better Buying Index Report* Spring 2018, betterbuy
ing.org, https://betterbuying.org/wp-content/uploads/2018/05/4159_better_buying
_report_final.pdf, 18.

102 **CSR policies are often:** Julia Bonner and Adam Friedman, "Corporate Social Responsibility: Who's Responsible?," Public Relations Society of America, n.d., https://
apps.prsa.org/intelligence/partnerresearch/partners/nyu_scps/corporatesocialrespon
sibility.pdf.

102 **the case of Macy's:** "Cheryl Heinonen Named Macy's, Inc. Executive Vice President, Corporate Communications," businesswire.com, December 19, 2016, https://
www.businesswire.com/news/home/20161219006227/en/Cheryl-Heinonen
-Named-Macys-Executive-Vice-President.

102 **At Walmart the person:** "Dan Bartlett, Executive Vice President, Corporate Affairs," Walmart.com, https://corporate.walmart.com/our-story/leadership/executive
-management/dan-bartlett/.

102 **"high pressure negotiating strategies":** *Better Buying Index Report* Spring 2018, 24.

102 **per-unit prices in Bangladesh:** Mohammad Nurul Alam, "Bangladesh-Made Garment Price Drops 1.61% in Last 4 Years," *TextileToday*, July 18, 2019, https://www
.textiletoday.com.bd/bangladesh-made-garment-price-drops-1-61-last-4-years/.

103 **delays create logistical nightmares:** "From Obligation to Opportunity: A Market Systems Analysis of Working Conditions in Asia's Garment Export Industry,"
September 2017, ILO.org, https://www.ilo.org/wcmsp5/groups/public/—ed_emp
/—emp_ent/—ifp_seed/documents/publication/wcms_628430.pdf, 47.

103 **"16 percent of buyers":** *Better Buying Index Report Spring* 2018, 24.

103 **52 percent of apparel suppliers:** Mengxin Li, "Paying for a Bus Ticket and Expecting to Fly: How Apparel Brand Purchasing Practices Drive Labor Abuses," Human
Rights Watch, 2019, 2, https://www.hrw.org/report/2019/04/24/paying-bus-ticket
-and-expecting-fly/how-apparel-brand-purchasing-practices-drive.

103 **occurring in Bangladesh:** Li, "Paying for a Bus Ticket and Expecting to Fly," 50.

103 **a "widespread practice":** Li, "Paying for a Bus Ticket and Expecting to Fly," 32.

103 **infected with the virus:** Fashion Revolution, "The Impact of COVID-19 on the
People Who Make Our Clothes," https://www.fashionrevolution.org/the-impact-of
-covid-19-on-the-people-who-make-our-clothes/.

103 **discounts on orders:** Amy Bainbridge and Supattra Vimonsuknopparat, "Suppliers Under Pressure as Australian Retailers Ask for Discounts, Hold Orders During
Coronavirus Pandemic," ABC.net.au, May 12, 2020, https://www.abc.net.au/news
/2020-05-13/australian-retailers-delay-supplier-payments-amid-coronavirus
/12236458.

104 **media and advocacy pressure:** "Major Apparel Brands Delay & Cancel Orders in Response to Pandemic, Risking Livelihoods of Millions of Garment Workers in Their Supply Chains," Business & Human Rights Resource Centre, March 24, 2020, https://www.business-humanrights.org/en/major-apparel-brands-delay-cancel -orders-in-response-to-pandemic-risking-livelihoods-of-millions-of-garment -workers-in-their-supply-chains.

105 **"Workers shall be paid":** "Code of Vendor Conduct," Gap Inc., revised June 2016, https://gapinc-prod.azureedge.net/gapmedia/gapcorporatesite/media/images /docs/codeofvendorconduct_final.pdf.

105 **verbiage that's typical:** "Toward Fair Compensation in Bangladesh," Fair Labor Association, April 2018, https://www.fairlabor.org/sites/default/files/documents/reports /toward_fair_compensation_in_bangladesh_april_2018_1.pdf; and Matt Cowgill and Phu Huynh, "Weak Minimum Wage Compliance in Asia's Garment Industry," ILO.org, August 2016, https://www.ilo.org/wcmsp5/groups/public/—ed_protect /—protrav/—travail/documents/publication/wcms_509532.pdf.

106 **wage-related violations were present:** "Better Work: Stage II Global Compliance Synthesis Report 2009–2012," The Better Work Global Programme, 2013, https:// betterwork.org/wp-content/uploads/2020/01/Global-Synthesis-Report-final.pdf.

106 **51 percent of Indian garment factories:** Cowgill and Huynh, "Weak Minimum Wage Compliance in Asia's Garment Industry," Appendix A.

106 **not a "living wage":** "Memorandum of Understanding," Action, Collaboration, Transformation, 2020, https://actonlivingwages.com/memorandum-of-understanding/.

106 **government-set minimum wages:** Clean Clothes Campaign, "Living Wage in Asia," Clean Clothes Campaign, 2014, https://archive.cleanclothes.org/resources /publications/asia-wage-report/view.

107 **aka "shadow" factories:** Andy Kroll, "Are Walmart's Chinese Factories as Bad as Apple's?," *Mother Jones*, March/April 2012, https://www.motherjones.com/envi ronment/2012/03/walmart-china-sustainability-shadow-factories-greenwash/.

107 **in Bangladesh estimate about half:** Li, "Paying for a Bus Ticket and Expecting to Fly," 6.

107 **top three garment-producing countries:** "U.S. Department of Labor's 2018 List of Goods Produced by Child Labor or Forced Labor," https://www.dol.gov/agencies /ilab/reports/child-labor/list-of-goods.

108 **New Deal benefits:** Jonathan Grossman, "Fair Labor Standards Act of 1938: Maximum Struggle for a Minimum Wage," U.S. Department of Labor, https://www .dol.gov/general/aboutdol/history/flsa1938.

108 **Industrial Revolution–era:** "History of Child Labor in the United States—Part 1: Little Children Working," *Monthly Labor Review*, U.S. Bureau of Labor Statistics, January 2017, https://www.bls.gov/opub/mlr/2017/article/history-of-child-labor-in -the-united-states-part-1.htm.

108 **current-day garment factory:** Josephine Moulds, "Child Labour in the Fashion Supply Chain," *Guardian*, https://labs.theguardian.com/unicef-child-labour/.

109 **illegal in China:** "Workers' Rights and Labour Relations in China," *China Labour Bulletin*, August 13, 2020, https://clb.org.hk/content/workers%E2%80%99-rights -and-labour-relations-china.

111 **fire in the Ha-Meem factory:** Brian Ross, Matthew Mosk, and Cindy Galli, "Workers Die at Factories Used by Tommy Hilfiger," ABC News, March 21, 2012, https://abcnews.go.com/Blotter/workers-die-factories-tommy-hilfiger/story?id=15966305.

111 **Hilfiger's parent company PVH:** Kevin Douglas Grant, "Tommy Hilfiger Caves on Factory Labor Conditions Ahead of ABC Report," *GlobalPost*, March 21, 2012, https://www.pri.org/stories/2012-03-21/tommy-hilfiger-caves-factory-labor-conditions-ahead-abc-report.

111 **an agreement that is the basis:** "The History Behind the Bangladesh Fire and Safety Accord," July 8, 2013, Clean Clothes Campaign, Maquila Solidarity Network, https://digitalcommons.ilr.cornell.edu/cgi/viewcontent.cgi?article=2844&context=globaldocs.

112 **the Accord has elevated:** "Accord on Fire and Building Safety in Bangladesh," May 13, 2013, bangladesh.wpengine.com, https://bangladesh.wpengine.com/wp-content/uploads/2018/08/2013-Accord.pdf.

112 **more than 1,600 factories:** Ritika Iyer, "Protecting the Safety of Bangladeshi Garment Workers," April 4, 2019, http://gppreview.com/2019/04/04/protecting-safety-bangladeshi-garment-workers/.

112 **eight domestic and two international:** "Frequently Asked Questions (FAQ) About the Bangladesh Safety Accord," Clean Clothes Campaign, https://cleanclothes.org/issues/faq-safety-accord#2—who-signed-the-accord-on-fire-and-building-safety-in-bangladesh-.

112 **more than 190 signatories:** "About," Accord on Fire and Building Safety in Bangladesh, https://bangladeshaccord.org/about/.

112 **the inspected factories' statuses:** "Factories," Accord on Fire and Building Safety in Bangladesh, https://bangladeshaccord.org/factories.

112 **a transition agreement was signed:** Monira Munni, "Let Accord Work Independently During Its Transition Period," *Financial Express*, June 3, 2019, https://thefinancialexpress.com.bd/trade/let-accord-work-independently-during-its-transition-period-1559579014.

112 **Readymade Sustainability Council:** Jasmin Malik Chua, "Readymade Sustainability Council to Take Over from Bangladesh Accord," *Sourcing Journal*, September 2019, https://sourcingjournal.com/topics/labor/readymade-sustainability-council-bangladesh-167523.

112 **they are *legally bound*:** "Covington Helps Secure Historic Settlement in Arbitration Under the Accord on Fire & Building Safety in Bangladesh," January 22, 2018, https://www.cov.com/en/news-and-insights/news/2018/01/covington-helps-secure-historic-settlement-in-arbitration-under-the-accord-on-fire-and-building-safety-in-bangladesh.

113 **50 percent of the board:** Benjamin A. Evans, "Accord on Fire and Building Safety in Bangladesh: An International Response to Bangladesh Labor Conditions," *North Carolina Journal of International Law* 45, no. 4 (2020): 598, https://core.ac.uk/download/pdf/151516597.pdf.

114 **"If you want markets":** Dani Rodrik, *The Globalization Paradox* (New York: W. W. Norton, 2011), 211.

115 **"The misery of being exploited":** Christopher Blattman and Stefan Dercon, "Everything We Knew About Sweatshops Was Wrong," *New York Times*, April 27, 2017,

https://www.nytimes.com/2017/04/27/opinion/do-sweatshops-lift-workers
-out-of-poverty.html.

CHAPTER 5: RECLAIMING ESSENTIALS FOR ALL:
PACKING AND DISTRIBUTION

118 **Laura started her career:** Interviewees' names in this chapter have been changed to protect their privacy.

119 **more than 7 percent:** Federal Register, Department of Health and Human Services, Administration for Children and Families, 67440, https://www.govinfo.gov/content/pkg/FR-2016-09-30/pdf/2016-22986.pdf.

119 **cost of childcare has risen:** Mark Lino et al., "Expenditures on Children by Families, 2015," U.S. Department of Agriculture, 21, https://fns-prod.azureedge.net/sites/default/files/crc2015_March2017.pdf.

119 **second-largest private employer:** Charles Duhigg, "Is Amazon Unstoppable?," *The New Yorker*, October 10, 2019, https://www.newyorker.com/magazine/2019/10/21/is-amazon-unstoppable.

119 **It only just snatched:** "Amazon Apparel: Annual US Survey Reveals Amazon Has Overtaken Walmart as America's Most-Shopped Retailer for Apparel," Coresight Research, March 4, 2019, https://coresight.com/research/amazon-apparel-annual-us-survey-reveals-amazon-has-overtaken-walmart-as-americas-most-shopped-retailer-for-apparel/. Amazon does not report separate fashion sales.

119 **online commerce beat out:** Kate Rooney, "Online Shopping Overtakes a Major Part of Retail for the First Time Ever," CNBC, April 2, 2019, https://www.cnbc.com/2019/04/02/online-shopping-officially-overtakes-brick-and-mortar-retail-for-the-first-time-ever.html.

120 **more than one third:** Chavie Lieber, "Will Fashion Ever Really Embrace Amazon?," Business of Fashion, February 28, 2020, https://www.businessoffashion.com/articles/professional/fashion-brands-how-to-sell-on-amazon.

120 **US online sales in June 2020:** Katie Evans, "More Than One-Third of Consumers Shop Online Weekly Since Coronavirus Hit," Digital Commerce 360, October 21, 2020, https://www.Digitalcommerce360.Com/Article/Coronavirus-Impact-Online-Retail/.

120 **these behavioral changes:** Marc Bain, "How Covid-19 Could Change Fashion and Retail, According to Experts," Quartz.com, https://qz.com/1831203/how-covid-19-could-change-fashion-and-retail/.

121 **Half of Amazon's private labels:** Don Davis, "Amazon Triples Its Private-Label Product Offerings in 2 Years," Digital Commerce 360, May 20, 2020, https://www.digitalcommerce360.com/2020/05/20/amazon-triples-its-private%E2%80%91label-product-offerings-in-2-years/.

121 *Vogue* x **Amazon Fashion:** Vanessa Friedman, "Amazon to the Rescue of the Fashion World!," *New York Times*, May 14, 2020, https://www.nytimes.com/2020/05/14/fashion/amazon-vogue-CFDA.html.

121 **"Other employers feel":** Shirin Ghaffary and Jason Del Rey, "The Real Cost of Amazon," *Vox*, June 29, 2020, https://www.vox.com/recode/2020/6/29/21303643

/amazon-coronavirus-warehouse-workers-protest-jeff-bezos-chris-smalls-boycott
-pandemic.

121 **250,000 such workers:** About Amazon Staff, "Fulfillment in Our Buildings," About
Amazon.com, https://www.aboutamazon.com/amazon-fulfillment/our-fulfillment
-centers/fulfillment-in-our-buildings.

121 **median annual salary:** Shira Ovide, "Amazon Is Defined by Billions and Millions; Median Salary Is $28,446," *Bloomberg*, April 19, 2018, https://www.bloom
berg.com/opinion/articles/2018-04-19/amazon-is-defined-by-billions-median
-salary-is-28-446.

121 **go on government assistance:** The Counter reported that one in three of Amazon's
employees in Arizona depend on SNAP benefits, while in Pennsylvania and Ohio,
the figure appears to be around one in ten. See H. Claire Brown, "Amazon Gets
Huge Subsidies to Provide Good Jobs—but It's a Top Employer of SNAP Recipi-
ents in at Least Five States," The Counter, April 18, 2018, https://thecounter.org
/Amazon-Snap-Employees-Five-States/.

121 **not pay a single cent:** Tom Huddleston Jr., "Amazon Had to Pay Federal Income
Taxes for the First Time Since 2016—Here's How Much," CNBC, February 4,
2020, https://www.cnbc.com/2020/02/04/amazon-had-to-pay-federal-income-taxes
-for-the-first-time-since-2016.htm.

122 **hitting $204.6 billion:** Jonathan Ponciano, "Jeff Bezos Becomes the First Person
Ever Worth $200 Billion," *Forbes*, August 26, 2020, https://www.forbes.com/sites
/jonathanponciano/2020/08/26/worlds-richest-billionaire-jeff-bezos-first-200
-billion/#5357fdbc4db7.

122 **substantially above Bill Gates:** Jonathan Ponciano, "The World's 10 Richest
Billionaires Lose $38 Billion on Coronavirus-Spurred 'Black Monday,'" *Forbes*,
March 9, 2020, https://www.forbes.com/sites/jonathanponciano/2020/03/09/the
-worlds-10-richest-people-lose-38-billion-on-coronavirus-spurred-black
-monday/#3063d91846ed.

122 **broke the $200 billion mark:** Amit Chowdry, "Net Worth Update: Jeff Bezos
Now over $200 Billion and Elon Musk over $100 Billion," Pulse 2.0, August 31,
2020, https://pulse2.com/jeff-bezos-elon-musk-net-worth/.

122 **"Amazon retail heroes":** "Thank You Amazon Heroes," iSpot.tv, 2020, https://
www.ispot.tv/ad/nkaN/amazon-covid-19-thank-you-amazon-heroes#.

122 **Other ad spots:** "Meet Janelle," iSpot.tv, 2020, https://www.ispot.tv/ad/nBRx
/amazon-meet-janelle.

123 **bonus of $500:** Ghaffary and Del Rey, "The Real Cost of Amazon."

123 **retail and online shopping:** "US Manufacturing Decline and the Rise of New
Production Innovation Paradigms," OECD, 2017, https://www.oecd.org/united
states/us-manufacturing-decline-and-the-rise-of-new-production-innovation
-paradigms.htm.

123 **at least $1 billion in incentives:** Janelle Jones and Ben Zipperer, "Unfulfilled
Promises," Economic Policy Institute, February 1, 2018, https://www.epi.org/pub
lication/unfulfilled-promises-amazon-warehouses-do-not-generate-broad-based
-employment-growth/.

124 **"Work hard. Have fun.":** Ghaffary and Del Rey, "The Real Cost of Amazon."

125 **but the associates:** Amazon refers to its distribution employees as "associates," a subtle linguistic argument that people working for the company are or should be deeply aligned with those who manage and own it.

125 **2,500 full-time employees:** Brent Johnson, "A Peek Inside Amazon's Robot-Filled Edison Facility (and Murphy's Update on HQ2 Bid)," NJ.com, May 14, 2019, https://www.nj.com/politics/2018/09/a_peek_inside_amazons_robot-filled_facility_in _edison_and_murphys_update_on_hq2.html.

126 **twenty hours of UTO:** Isobel Asher Hamilton, "'It's a Slap in the Face': Amazon Is Handing Out 'Thank You' T-Shirts to Warehouse Workers as It Cuts Their Hazard Pay," *Business Insider,* May 16, 2020, https://www.businessinsider.com/amazon -warehouse-workers-thank-you-t-shirts-as-it-cuts-their-hazard-pay-2020-5.

127 **26 use these:** About Amazon Staff, "What Robots Do (and Don't Do) at Amazon Fulfillment Centers," AboutAmazon, https://www.aboutamazon.com/amazon-ful fillment/our-innovation/what-robots-do-and-dont-do-at-amazon-fulfillment -centers/.

130 **Jeff Bezos's net worth increased:** Jack Pitcher, "Jeff Bezos Adds Record $13 Billion in Single Day to Fortune," *Bloomberg,* July 20, 2020, https://www.bloomberg.com /news/articles/2020-07-20/jeff-bezos-adds-record-13-billion-in-single-day-to-his -fortune?sref=1gzfmHYv.

131 **adding these 200,000 robots:** PYMNTS, "Amazon Has Used Over 200,000 Robotic Drives Around the World," PYMNTS.com, June 5, 2019, https://www .pymnts.com/amazon/2019/robotic-warehouse-automation/; and "Amazon Empire": Jeff Wilke interview, *Frontline,* PBS, February 18, 2020, https://www.youtube .com/watch?v=hziCY1ohf64.

131 **to meet faster fulfillment:** "Amazon Empire": Jeff Wilke interview.

131 **a new German company:** Adam Satariano and Cade Metz, "A Warehouse Robot Learns to Sort Out the Tricky Stuff," *New York Times,* January 29, 2020, https:// www.nytimes.com/2020/01/29/technology/warehouse-robot.html.

134 **jobs without a degree of control:** Michael Marmot, *Status Syndrome: How Your Place on the Social Gradient Affects Your Health* (London: Bloomsbury, 2004), 130.

134 **"You aren't a machine":** Johann Hari, *Lost Connections: Uncovering the Real Causes of Depression* (New York: Bloomsbury, 2018), 308.

136 **average turnover rate:** Irene Tung and Deborah Berkowitz, "Amazon's Disposable Workers: High Injury and Turnover Rates at Fulfillment Centers in California," National Employment Law Project, March 6, 2020, https://www.nelp.org/publi cation/amazons-disposable-workers-high-injury-turnover-rates-fulfillment -centers-california.

137 **Great Affordability Crisis:** Annie Lowrey, "The Great Affordability Crisis Breaking America," *Atlantic,* February 7, 2020, https://www.theatlantic.com/ideas/archive /2020/02/great-affordability-crisis-breaking-america/606046.

137 **it would be $24/hour:** "The Productivity-Pay Gap," Economic Policy Institute, updated July 2019, https://www.epi.org/productivity-pay-gap/.

137 **17 percent of adults:** "Report on the Economic Well-Being of U.S. Households in 2018," Federal Reserve, May 2019, https://www.federalreserve.gov/publications /files/2018-report-economic-well-being-us-households-201905.pdf.

137 **$400 in savings:** Kirsty Bowen, "Employers Pay When Workers Face Financial Precarity," Futurity, December 2, 2019, https://www.futurity.org/financial-precarity -money-worries-middle-class-employees-2222922-2/.

137 **America's top earners:** Raj Chetty et al., "The Association Between Income and Life Expectancy in the United States, 2001–2014," National Institutes of Health, 2016, https://www.ncbi.nlm.nih.gov/pmc/articles/PMC4866586/ https:// healthinequality.org. See also *Journal of the American Medical Association* 317, no. 1 (January 3, 2017): 90.

137 **Americans were on food stamps:** "How the Share of Americans Receiving Food Stamps Has Changed," USA Facts, https://usafacts.org/articles/snap-benefits-how -share-americans-receiving-food-stamps-has-changed/.

137 **twice as many billionaires:** Reuters, "We're in the Longest Economic Expansion Ever—but It's the Rich Who Are Getting Richer," NBC News, July 2, 2019, https:// www.nbcnews.com/business/economy/we-re-longest-economic-expansion -ever-it-s-rich-who-n1025611.

138 **CEO earned $924,000:** Lawrence Mishel and Julia Wolfe, "CEO Compensation Has Grown 940% Since 1978," Economic Policy Institute, August 14, 2019, https:// www.epi.org/publication/ceo-compensation-2018/.

138 **top 1 percent of earners:** Eric Levitz, "The One Percent Have Gotten $21 Trillion Richer Since 1989. The Bottom 50% Have Gotten Poorer," *New York* magazine, June 16, 2019, http://nymag.com/intelligencer/amp/2019/06/the-fed-just-released-a -damning-indictment-of-capitalism.html.

138 **creating the most profit:** Cydney Posner, "So Long to Shareholder Primacy," Harvard Law School Forum on Corporate Governance, August 22, 2019, https://cor pgov.law.harvard.edu/2019/08/22/so-long-to-shareholder-primacy/.

138 **The better the stock:** Michael C. Jensen and Kevin J. Murphy, "CEO Incentives— It's Not How Much You Pay, But How," *Harvard Business Review*, May–June 1990, https://hbr.org/1990/05/ceo-incentives-its-not-how-much-you-pay-but-how.

138 **above-market compensation:** Robin Ferracone, "Dare to Be Different—The Case of Amazon.com," *Forbes*, April 23, 2019, https://www.forbes.com/sites/robinfer racone/2019/04/23/dare-to-be-different-the-case-of-amazon-com/#139712 f6cb99.

138 **colonizing outer space:** Isobel Asher Hamilton, "Jeff Bezos Took Another Veiled Shot at Elon Musk, Arguing That Reaching Mars Is an 'Illusion' Without Going Via the Moon," *Business Insider*, June 20, 2019, https://www.businessinsider.com /jeff-bezos-elon-musk-must-go-to-moon-before-mars-2019-6.

138 **decreased every year since 2014:** Life expectancy plateaued in 2011, decreased from 2014 to 2018, went up from 2018 to 2019. See Kaitlin Sullivan, "U.S. Life Expectancy Goes Up for the First Time Since 2014," NBC News, January 30, 2020, https://www.nbcnews.com/health/health-news/u-s-life-expectancy-goes-first -time-2014-n1125776.

138 *lower average* **life expectancies:** U.S. Burden of Disease Collaborators, "The State of US Health, 1990–2016, Burden of Disease, Injuries, and Risk Factors Among US States," JAMA Network, April 10, 2018, https://jamanetwork.com/journals /jama/fullarticle/2678018; compared to Bangladesh, from WHO data, "Life Ex-

pectancy and Healthy Life Expectancy, Data by Country," available at https://apps
.who.int/gho/data/node.main.688?lang=en.

138 **"Rising economic and political power"**: Zachary Siegel, "Capitalism Is Killing
Us," *The Nation*, April 23, 2020, https://www.thenation.com/article/culture/case
-deaton-deaths-of-despair-book-review/.

140 **workers getting fired**: Tamar Lapin, "Amazon Fires Organizer of Strike at Staten
Island Warehouse," *New York Post*, March 30, 2020, available at https://www.mar
ketwatch.com/story/amazon-fires-organizer-of-strike-at-staten-island-warehouse
-2020-03-30.

140 **monitoring the union-organizing**: Lauren Kaori Gurley, "Secret Amazon Reports
Expose the Company's Surveillance of Labor and Environmental Groups," *Vice*,
November 23, 2020, available at https://www.vice.com/en/article/5dp3yn/amazon
-leaked-reports-expose-spying-warehouse-workers-labor-union-environmental
-groups-social-movements.

140 **"not just due to COVID-19"**: Ghaffary and Del Rey, "The Real Cost of Amazon."

140 **Union members earn**: Lawrence Mishel, "Unions, Inequality, and Faltering Middle-
Class Wages," Economic Policy Institute, August 29, 2012, https://www.epi.org/pub
lication/ib342-unions-inequality-faltering-middle-class/.

141 **"unionization is likely"**: Jason Del Rey and Shirin Ghaffary, "Amazon White-Collar
Employees Are Fuming over Management Targeting a Fired Warehouse Worker,"
Vox, April 5, 2020, https://www.vox.com/recode/2020/4/5/21206385/amazon-fired
-warehouse-worker-christian-smalls-employee-backlash-david-zapolsky-coronavirus.

141 **unionized Black workers**: Cherrie Bucknor, "Black Workers, Unions, and Inequality," Center for Economic and Policy Research, August 2016, https://cepr.net/im
ages/stories/reports/black-workers-unions-2016-08.pdf?v=2.

141 **significant percentage of Black employees**: Karen Weise, "Amazon Workers
Urge Bezos to Match His Words on Race with Actions," *New York Times*, June 24,
2020, https://www.nytimes.com/2020/06/24/technology/amazon-racial-inequality
.html.

141 **13.4 percent Black**: US Census Bureau, Quick Facts, https://www.census.gov
/quickfacts/fact/table/US/PST045219.

141 **overrepresentation of Black workers**: Danyelle Solomon, Connor Maxwell, and
Abril Castro, "Systematic Inequality and Economic Opportunity," Center for American Progress, August 7, 2019, https://www.americanprogress.org/issues/race/reports
/2019/08/07/472910/systematic-inequality-economic-opportunity/; and "Southern
Black Codes," Constitutional Rights Foundation, https://www.crf-usa.org/brown-v
-board-50th-anniversary/southern-black-codes.html.

141 **Jim Crow laws**: Juan F. Perea, "The Echoes of Slavery: Recognizing the Racist
Origins of the Agricultural and Domestic Worker Exclusion from the National
Labor Relations Act," Loyola University Chicago, School of Law, 2011, https://
lawcommons.luc.edu/cgi/viewcontent.cgi?article=1150&context=facpubs.

141 **redlining allowed housing discrimination**: Andre M. Perry and David Harsh-
barger, "America's Formerly Redlined Neighborhoods Have Changed, and So Must
Solutions to Rectify Them," October 14, 2019, https://www.brookings.edu/research
/americas-formerly-redlines-areas-changed-so-must-solutions/.

141 **its impact continues:** Cory Turner et al., "Why America's Schools Have a Money Problem," NPR, April 18, 2016, https://www.npr.org/2016/04/18/474256366/why -americas-schools-have-a-money-problem.

141 **de facto segregation and underfunding:** This is a grossly cursory overview of the wide-reaching impact of racism on our country's institutions and attitudes. It is an area that is in profound need of greater understanding and reflection. The works of Angela Y. Davis, Steven A. Reich, Nick Estes, Mehrsa Baradaran, and Resmaa Menakem have all been helpful to me and I recommend them highly.

142 **Black people also took a harder hit:** Mark Muro, Robert Maxim, and Jacob Whiton, "Automation and Artificial Intelligence," Brookings Institution, January 2019, 45–46, https://www.brookings.edu/wp-content/uploads/2019/01/2019.01 _BrookingsMetro_Automation-AI_Report_Muro-Maxim-Whiton-FINAL -version.pdf.

142 **Since 1954 Black workers suffered:** Drew Desilver, "Black Unemployment Rate Is Consistently Twice That of Whites," Pew Research Center Fact Tank, August 21, 2013, https://www.pewresearch.org/fact-tank/2013/08/21/through-good-times-and -bad-black-unemployment-is-consistently-double-that-of-whites/.

142 **the lowest-paying occupations:** Edward Rodrigue and Richard V. Reeves, "Five Bleak Facts on Black Opportunity," Brookings Institution, January 15, 2015, https://www .brookings.edu/blog/social-mobility-memos/2015/01/15/five-bleak-facts-on-black -opportunity/.

142 **for upward mobility:** Jeanna Smialek and Jim Tankersley, "Black Workers, Already Lagging, Face Big Economic Risks," *New York Times*, April 2020, https:// www.nytimes.com/2020/06/01/business/economy/black-workers-inequality -economic-risks.html,=.

142 **"Our needs are identical":** Staff Writer, "Letter: It's Really This Simple," *Panama City News Herald*, June 24, 2020, https://www.newsherald.com/story/opinion/letters /2020/06/24/letter-itrsquos-really-this-simple/41957943/.

142 **Isabel Wilkerson convincingly describes:** Terry Gross, "It's More Than Racism: Isabel Wilkerson Explains America's 'Caste' System," *Fresh Air*, NPR, August 4, 2020, https://www.npr.org/2020/08/04/898574852/its-more-than-racism-isabel -wilkerson-explains-america-s-caste-system.

142 **more than $16 million on lobbying:** "Client Profile: Amazon.com," OpenSecrets .org, https://www.opensecrets.org/federal-lobbying/clients/summary?cycle=2019&id =D000023883.

142 **increased by almost 470 percent:** Franklin Foer, "Jeff Bezos's Master Plan," *Atlantic*, November 2019, https://www.theatlantic.com/magazine/archive/2019/11/what -jeff-bezos-wants/598363/.

143 **spend the most on lobbying:** Lee Drutman, "How Corporate Lobbyists Conquered American Democracy," *Atlantic*, April 20, 2015, https://www.theatlantic .com/business/archive/2015/04/how-corporate-lobbyists-conquered-american -democracy/390822/.

143 **hiring government officials:** Foer, "Jeff Bezos's Master Plan."

143 **countries with more collective bargaining:** OECD Employment Outlook 2018, chapter 3, https://www.oecd-ilibrary.org/sites/empl_outlook-2018-7-en/index.html ?itemId=/content/component/empl_outlook-2018-7-en.

144 **develop nationwide plans:** World Investment Report, 2018, chapter 4, https://
unctad.org/en/PublicationChapters/wir2018ch4_en.pdf.

144 **New Deal policies:** Michael Tomasky, "Unemployment in the 30s: The Real
Story," *Guardian,* February 9, 2009, https://www.theguardian.com/commentisfree
/michaeltomasky/2009/feb/09/obama-administration-usemployment-new-deal
-worked.

145 **the phrase "good jobs":** Dani Rodrik and Charles Sabel, "Building a Good Jobs
Economy," November 2019, HKS Working Paper No. RWP20-001, available at
SSRN: https://ssrn.com/abstract=3533430 or http://dx.doi.org/10.2139/ssrn.3533430.

145 **vote for authoritarian values:** Rodrik and Sabel, "Building a Good Jobs Economy."

145 **negatively impacted by globalization:** Rodrik and Sabel, "Building a Good Jobs
Economy."

146 **"I think one thing":** Susan Glasser and Glenn Thrush, "What's Going On with
America's White People?," *Politico,* September/October 2016, https://www.politico
.com/magazine/story/2016/09/problems-white-people-america-society-class-race
-214227.

146 **Revamping our economy:** Rodrik and Sabel, "Building a Good Jobs Economy."

147 **"The long history of capitalism":** "Sven Beckert on Inequality, Jobs, and Capitalism,"
Harvard Kennedy School Mossavar-Rahmani Center for Business and Government,
August 2020, https://www.hks.harvard.edu/centers/mrcbg/programs/growthpolicy
/sven-beckert-inequality-jobs-and-capitalism.

CHAPTER 6: MORE IS MORE: CONSUMERISM GOES VIRAL

150 **971 square feet:** "2017 Characteristics of New Housing," U.S. Department of
Commerce, https://www.census.gov/construction/chars/pdf/c25ann2017.pdf.

150 **pandemic job loss in April 2020:** "Databases, Tables & Calculators by Subject:
Labor Force Statistics from the Current Population Survey," U.S. Bureau of Labor
Statistics, October 15, 2020, https://data.bls.gov/timeseries/LNS14000000.

150 **disproportionately Black and Latino:** "State Unemployment by Race and Ethnic-
ity," Economic Policy Institute, updated August 2020, https://www.epi.org/indi
cators/state-unemployment-race-ethnicity/.

150 **10.8 percent for white workers:** "The Employment Situation—September 2020,"
Bureau of Labor Statistics News Release, https://www.bls.gov/news.release/pdf
/empsit.pdf; and "Labor Force Statistics from the Current Population Survey: Household Data," U.S. Bureau of Labor Statistics, https://www.bls.gov/web/empsit/cpsee
_e16.htm.

151 **disposable fashion machine:** Anna Granskog et al., "Survey: Consumer Sentiment
on Sustainability in Fashion," McKinsey, https://www.mckinsey.com/industries
/retail/our-insights/survey-consumer-sentiment-on-sustainability-in-fashion.

153 **a member of the French court:** Kimberly Chrisman-Campbell, "The King of Cou-
ture," *Atlantic,* September 1, 2015, https://www.theatlantic.com/entertainment/ar
chive/2015/09/the-king-of-couture/402952//. Perhaps not surprisingly, the whole
idea of the fashion season is coming into question under COVID-19; according to

the McKinsey & Company survey, 65 percent of respondents supported postponing the launch of collections, 58 percent of respondents said they were not as interested in the "fashion clothing" because of the pandemic, and "newness" became the least important factor to their purchases; https://www.mckinsey.com/industries/retail /our-insights/survey-consumer-sentiment-on-sustainability-in-fashion.

153 **"When I came back"**: Josef Seethaler, ed., *Selling War: The Role of the Mass Media in Hostile Conflicts from World War I to the "War on Terror"* (Chicago: University of Chicago Press/Intellect Books, 2013), 111.

155 **"It's not that you think"**: *The Century of Self,* directed by Adam Curtis (London: BBC, 2002).

155 **every eleven-odd seconds**: Simone Stolzoff, "Jeff Bezos Will Still Make the Annual Salary of His Lowest-Paid Employees Every 11.5 Seconds," Quartz.com, October 2, 2018, https://qz.com/work/1410621/jeff-bezos-makes-more-than-his-least-amazon -paid-worker-in-11-5-seconds/.

157 **"the American citizen's"**: Samuel Strauss, "Things Are in the Saddle," *Atlantic,* November 1924, https://memory.loc.gov/cgi-bin/ampage?collID=cool&item Link=r?ammem/coolbib:@field(NUMBER+@band(amrlgs+at1))&hdl=amrlgs :at1:002/.

157 **"It's not that the people"**: See *The Century of the Self—Part 1: "Happiness Machines,"* and Pema Levy, "The Secret to Beating Trump Lies with You and Your Friends," *Mother Jones,* November–December 2020, https://www.motherjones.com/politics /2020/10/relational-organizing.

157 **robust war machine**: *The War,* "War Production," PBS, https://www.pbs.org/thewar /at_home_war_production.htm.

158 **healthy middle class**: Jim Hightower, "The Rise and Fall of America's Middle Class," OtherWords.org, https://otherwords.org/rise-fall-americas-middle-class/.

158 **"We must shift America"**: *The Century of Self.*

158 **ten thousand advertisements a day**: Rex Briggs, "The Secret to Reaching Your Audience? Re-Evaluate What You're Paying For," *Adweek,* June 29, 2018, https://www .adweek.com/brand-marketing/the-secret-to-reaching-your-audience-re-evaluate -what-youre-paying-for/.

158 **they were exposed to around five hundred**: Caitlin Johnson, "Cutting Through Advertising Clutter," CBS *Sunday Morning,* September 1, 2006, https://www.cbsnews .com/news/cutting-through-advertising-clutter/.

158 **ads plays trickery**: Sophie Benson, "One & Done: Why Do People Ditch Their Clothes After Just One Wear?," Refinery29, October 3, 2019, https://www.refinery 29.com/en-gb/instagram-outfits-wear-once.

159 **"a set of values"**: Tim Kasser, *The High Price of Materialism* (Cambridge, MA: MIT Press, 2003), 489.

162 **getting products placed in movies**: Larry Tye, *The Father of Spin* (London: Picador, 2002), 59, https://www.google.com/books/edition/The_Father_of_Spin/GarJLYMm 3A0C?hl=en&gbpv=0.

163 **"When we win"**: Foer, "Jeff Bezos's Master Plan."

163 **Julia Roberts makes $50 million**: Brittany Irvine, "10 of the Highest Paid Celebrity Beauty Campaigns," Stylecaster, June 17, 2013, https://stylecaster.com/beauty/high est-paid-celebrity-beauty-campaigns/slide10#autoplay.

163 **between $100,000 and $250,000**: Aly Weisman, "Here's How Much Celebrities Are Paid to Wear Designer Dresses on the Red Carpet," *Business Insider*, June 15, 2015, https://www.businessinsider.com/how-much-celebrities-paid-to-wear-dresses-on-red-carpet-2015-6.

165 **"there's only so much"**: Alex Williams, "Why Don't Rich People Just Stop Working?," *New York Times*, October 17, 2019, https://www.nytimes.com/2019/10/17/style/rich-people-things.html.

165 **social media use grew 10 percent**: Simon Kemp, "Digital 2020: July Global Statshot," Datareportal, July 21, 2020, https://datareportal.com/reports/digital-2020-july-global-statshot.

165 **decreasing social media use**: Rimma Kats, "Time Spent on Social Media Is Anticipated to Increase 8.8%, Despite Expected Plateau," Emarketer.com, April 29, 2020, https://www.emarketer.com/content/us-adults-are-spending-more-time-on-social-media-during-the-pandemic.

166 **tens of thousands of dollars**: Amanda Perelli, "The Top 15 Makeup and Beauty YouTubers in the World, Some of Whom Are Making Millions of Dollars," *Business Insider,* August 6, 2019, https://www.businessinsider.com/beauty-and-makeup-youtube-channels-with-most-subscribers-2019-7/.

166 **$5 million**: Abram Brown, "TikTok's 7 Highest-Earning Stars: New Forbes List Led by Teen Queens Addison Rae and Charli D'Amelio," *Forbes*, August 6, 2020, https://www.forbes.com/sites/abrambrown/2020/08/06/tiktoks-highest-earning-stars-teen-queens-addison-rae-and-charli-damelio-rule/#373c4b075087.

166 **"Brands just send me"**: "The Real Story of Paris Hilton: This Is Paris Official Documentary," September 13, 2020, https://www.youtube.com/watch?v=wOg0TY1jG3w.

167 **86 percent of brands**: "The State of Fashion 2020," McKinsey & Company, https://www.mckinsey.com/~/media/McKinsey/Industries/Retail/Our%20Insights/The%20state%20of%20fashion%202020%20Navigating%20uncertainty/The-State-of-Fashion-2020-final.ashx, 34.

167 **Influencers also come at**: Chavie Lieber, "How and Why Do Influencers Make So Much Money? The Head of an Influencer Agency Explains," *Vox*, November 28, 2018, https://www.vox.com/the-goods/2018/11/28/18116875/influencer-marketing-social-media-engagement-instagram-youtube.

167 **Clothing is one of the top**: Jacques Bughin, "Getting a Sharper Picture of Social Media's Influence," McKinsey & Company, July 1, 2015, https://www.mckinsey.com/business-functions/marketing-and-sales/our-insights/getting-a-sharper-picture-of-social-medias-influence.

167 **Boohoo was growing**: Elizabeth Paton, "Why You Should Care That Boohoo Is Making Headlines This Week," *New York Times*, July 8, 2020, https://www.nytimes.com/2020/07/08/fashion/boohoo-labor-influencer-crisis.html.

168 **sales continued to grow**: Sarah Butler, "Boohoo Reports Sales Surge Despite Leicester Supplier Scandal," *Guardian*, September 30, 2020, https://www.theguardian.com/business/2020/sep/30/boohoo-reports-sales-surge-despite-leicester-supplier-scandal-covid.

169 **a Barnardos survey**: "Barley and Barnardo's—The Fast Fashion Crisis," Censuswide, https://censuswide.com/censuswide-projects/barley-and-barnardos-the-fast-fashion-crisis-research/.

169 **Barclay revealed that 9 percent:** "Snap and Send Back," Barclaycard, October 15, 2020, https://webcache.googleusercontent.com/search?q=cache:67aGeMNPgpoJ :https://www.home.barclaycard/media-centre/press-releases/snap-and-send-back .html+&cd=1&hl=en&ct=clnk&gl=us.

169 **the single-use garment:** Elizabeth Cline, "How Sustainable Is Renting Your Clothes, Really?," *Elle*, October 22, 2019, https://www.elle.com/fashion/a29536207 /rental-fashion-sustainability/.

170 **Over 80 percent:** "Mobile Fact Sheet," Pew Research Center, June 12, 2019, https://www.pewresearch.org/internet/fact-sheet/mobile/.

170 **half of all adults:** "How Many Smartphones Are in the World," bankmycell.com, https://www.bankmycell.com/blog/how-many-phones-are-in-the-world.

170 **ninety-six times per day:** "Americans Check Their Phones 96 Times a Day," Asurion, https://www.asurion.com/about/press-releases/americans-check-their-phones -96-times-a-day/.

170 **two hours shopping online:** "After the Binge the Hangover," Greenpeace, May 8, 2017, https://www.greenpeace.org/international/publication/6884/after-the-binge -the-hangover/.

170 **Forty-three percent of participants:** Granskog et al., "Survey: Consumer Sentiment on Sustainability in Fashion."

171 **kids as young as four:** Kasser, *The High Price of Materialism*, 499.

171 **negative effects of cell phone use:** Kasser, *The High Price of Materialism*, 495–96.

171 **Just having it nearby:** Henry H. Wilmer, Lauren E. Sherman, and Jason M. Chein, "Smartphones and Cognition: A Review of Research Exploring the Links Between Mobile Technology Habits and Cognitive Functioning," *Frontiers in Psychology* 8, no. 605 (2017): 4, https://www.frontiersin.org/articles/10.3389/fpsyg.2017.00605/full.

171 **frequent media multitaskers:** Wilmer, Sherman, and Chein, "Smartphones and Cognition," 6.

172 **that 95 percent of American teenagers:** Monica Anderson and Jingjing Jiang, "Teens, Social Media & Technology 2018," Pew Research Center, May 31, 2018, https://www.pewresearch.org/internet/2018/05/31/teens-social-media -technology-2018/.

172 **nine hours a day:** "Landmark Report: U.S. Teens Use an Average of Nine Hours of Media per Day, Tweens Use Six Hours," Common Sense Media, November 3, 2015, https://www.commonsensemedia.org/about-us/news/press-releases/landmark -report-us-teens-use-an-average-of-nine-hours-of-media-per-day.

172 **rate of suicide among girls:** Sally C. Curtin and Melonie Heron, "Death Rates Due to Suicide and Homicide Among Persons Aged 10–24: United States, 2000–2017," Centers for Disease Control, NCHS Data Brief No. 352, October 2019, https:// www.cdc.gov/nchs/data/databriefs/db352-h.pdf.

172 **cortisol levels:** Darby E. Saxbe and Rena Repetti, "For Better or Worse? Coregulation of Couples' Cortisol Levels and Mood States," *Journal of Personality and Social Psychology*, 98(1), 92-103 (2010), https://psycnet.apa.org/record/2009-24670-014.

172 **use their garages:** Jack Feuer, "The Clutter Culture," *UCLA Magazine*, July 1, 2012, http://magazine.ucla.edu/features/the-clutter-culture/index1.html.

172 **couples' levels of cortisol:** Darby E. Saxbe and Rena Repetti, "No Place Like Home: Home Tours Correlate with Daily Patterns of Mood and Cortisol," *Person-

ality and Social Psychology Bulletin 36, no. 71 (November 23, 2009), https://dornsife
.usc.edu/assets/sites/496/docs/pubs/2009noplace.pdf.

173 **dopamine-pleasure spike:** M. Lenoir et al., "Intense Sweetness Surpasses Cocaine
Reward," *PLoS One* (August 1, 2007): 2:e698, https://journals.plos.org/plosone/article
?id=10.1371/journal.pone.0000698.

173 **rise of anxiety and depression:** "Depression: A Global Crisis," World Federation
for Mental Health, World Mental Health Day, October 10, 2020, https://www
.who.int/mental_health/management/depression/wfmh_paper_depression
_wmhd_2012.pdf.

174 **$689 million worth:** "Online Retailers Are Destroying Goods but Won't Say How
Much Ends Up as Trash," DW.com, n.d., https://www.dw.com/en/destroy-packages
-online-shopping/a-52281567.

174 **About 11 percent:** "US Retail: Many Unhappy Returns," *Financial Times,* January 5, 2020, https://www.ft.com/content/f92d875b-c2ea-4580-9c70-5e375f76dc6b
?accessToken=zwAAAW98oXJokdP5LYdbwupFgNOccF43X3bcaw.MEUCIQ
Codw25YOSv86v7fzE3P7SzNRRWHEwpXZzpvawm64IC_gIgc5QxjzCc3B-
d1xmuvD4wMlBq3PTnXDIj8DAM8jesGrho&sharetype=gift?token=ce8b22c4
-f4db-4899-8968-27068eb618e6.

174 **get thrown away:** "'It's Pretty Staggering,' Returned Online Purchases Often Sent
to Landfill Journalist's Research Reveals," CBC Radio, December 12, 2019, https://
www.cbc.ca/radio/thecurrent/the-current-for-dec-12-2019-1.5393783/it-s-pretty
-staggering-returned-online-purchases-often-sent-to-landfill-journalist-s
-research-reveals-1.5393806.

175 **doctors recommended their brand:** Martha N. Gardner and Allan M. Brandt,
"'The Doctors' Choice Is America's Choice': The Physician in US Cigarette Advertisements, 1930–1953," National Institutes of Health, https://www.ncbi.nlm.nih
.gov/pmc/articles/PMC1470496/; see also *American Journal of Public Health* 96,
no. 2 (February 2006): 222–32.

175 **The US surgeon general:** Jeffrey K. Stine, "Smoke Gets in Your Eyes: 20th Century
Tobacco Advertisements," National Museum of American History, March 17, 2014,
https://americanhistory.si.edu/blog/2014/03/smoke-gets-in-your-eyes-20th
-century-tobacco-advertisements.html.

175 **"Surgeon General's Warning":** "The 1964 Report on Smoking and Health,"
https://profiles.nlm.nih.gov/spotlight/nn/feature/smoking.

175 **America's smoking rate:** "50 Years of Progress Halves Smoking Rate, but Can
We Reach Zero?," NBC News, January 11, 2014, https://www.nbcnews.com/health
/cancer/50-years-progress-halves-smoking-rate-can-we-reach-zero-n7621.

176 **"Personal style, not fashion":** Rachel Tashjian, "The Most Sustainable Idea in
Fashion Is Personal Style," GQ.com, February 17, 2020, https://www.gq.com/story
/sustainable-fashion-personal-style.

178 **the idea of "spiritual consumerism":** Amanda Hess, "The New Spiritual Consumerism," *New York Times,* August 19, 2019, https://www.nytimes.com/2019/08/19/arts
/queer-eye-kondo-makeover.html?login=smartlock&auth=login-smartlock.

178 **consider "marketplace feminism":** Marcie Bianco, "We Sold Feminism to the
Masses, and Now It Means Nothing," Quartz.com, https://qz.com/692535/we-sold
-feminism-to-the-masses-and-now-it-means-nothing/.

CHAPTER 7: TIDYING UP: WHAT HAPPENS
TO CLOTHES WHEN WE GET RID OF THEM

180 **eighty pounds of textiles:** "Advancing Sustainable Materials Management: 2014 Fact Sheet," U.S. Environmental Protection Agency, November 2016, https://www .epa.gov/sites/production/files/2016-11/documents/2014_smmfactsheet_508.pdf.

180 **clothing and shoes:** "Nondurable Goods: Product-Specific Data," Facts and Figures About Materials, Waste and Recycling, U.S. Environmental Protection Agency, January 19, 2017, https://www.epa.gov/facts-and-figures-about-materials-waste-and -recycling/nondurable-goods-product-specific-data#ClothingandFootwear.

183 **200,000 tons of clothing:** https://www1.nyc.gov/assets/dsny/site/services/donate -goods/textiles.

184 **costs approximately $2.3 billion:** "12 Things New Yorkers Should Know About Their Garbage," Citizens Budget Commission, May 21, 2014, https://cbcny.org /research/12-things-new-yorkers-should-know-about-their-garbage.

185 **A full 56 percent:** "How Much of the City's Curbside Recyclables Get Properly Recycled," Independent Budget Office of the City of New York, New York City by the Numbers, July 14, 2016, https://ibo.nyc.ny.us/cgi-park2/2016/07/how-much-of -the-citys-curbside-recyclables-get-properly-recycled/.

185 **the right container:** Sally Goldenberg and Danielle Muoio, "Wasted Potential: Recycling Progress in Public Housing Eludes City Officials," *Politico*, January 7, 2020, https://www.politico.com/states/new-york/city-hall/story/2020/01/07/wasted -potential-recycling-progress-in-public-housing-eludes-city-officials-1246328.

185 **$106 million in budget cuts to DSNY:** Clodagh McGowan, "How the City's Budget Cuts Will Impact Sanitation Operations," NY1.com, July 7, 2020, https:// www.ny1.com/nyc/all-boroughs/news/2020/07/07/dsny-sanitation-budget-cuts -what-to-expect-.

185 **Sunday basket trash pickup:** Sydney Pereira, "Facing Criticism over Pile-Ups, De Blasio Will Restore Some Garbage Collection Services," Gothamist, September 16, 2020, https://gothamist.com/news/facing-criticism-over-pile-ups-de-blasio-will -restore-some-garbage-collection-services.

186 **sanitation workers are frontline heroes:** Adam Minter, "Garbage Workers Are on the Virus Front Lines, Too," *Bloomberg*, March 23, 2020, https://www.bloomberg .com/opinion/articles/2020-03-23/coronavirus-outbreak-is-challenge-to-garbage -worker-safety.

186 **24,000 tons per day:** "How Much Garbage Does New York City Produce Daily? Tons," Metro.us, August 24, 2017, https://www.metro.us/how-much-garbage-does -new-york-city-produce-daily-tons/.

188 **3 million tons:** Steven Cohen, Hayley Martinez, and Alix Schroder, "Waste Management Practices in New York City, Hong Kong and Beijing," Columbia.edu, December 2015, http://www.columbia.edu/~sc32/documents/ALEP%20Waste%20 Managent%20FINAL.pdf.

188 **46.1 pounds of textile trash:** Interview with Tiffany Fuller, deputy director of the New York Department of Sanitation, February 12, 2020.

189 *Fortune* 500: "Waste Management, Fortune 500 #207," *Fortune*, May 18, 2020, https://fortune.com/fortune500/2019/waste-management/.

189 the city spent $411 million: "Report of the Finance Division on the Fiscal 2019 Preliminary Budget and the Fiscal 2018 Preliminary Mayor's Management Report for the Department of Sanitation," March 14, 2018, https://council.nyc.gov/budget /wp-content/uploads/sites/54/2018/03/FY19-Department-of-Sanitation.pdf.

189 plus another $87 million: "Report of the Finance Division on the Fiscal 2019 Preliminary Budget."

189 up from $300 million: New York City Independent Budget Office, "Waste Export Costs to Rise as Remaining Marine Transfer Stations Open," March 2017, https:// ibo.nyc.ny.us/iboreports/waste-export-costs-to-rise-as-remaining-marine-transfer -stations-open-march-2017.pdf.

189 increase to $420 million: Lisa M. Collins, "The Pros and Cons of New York's Fledgling Compost Program," *New York Times*, November 9, 2018, https://www .nytimes.com/2018/11/09/nyregion/nyc-compost-zero-waste-program.html.

189 $202 per ton for disposal: City of New York, "Mayor's Management Report, Preliminary Fiscal 2020," January 2020, https://www1.nyc.gov/assets/operations /downloads/pdf/pmmr2020/2020_pmmr.pdf.

189 more than 3 million tons of garbage: City of New York, "Mayor's Management Report," 104.

190 Race correlates to one's proximity: Robert D. Bullard et al., "Toxic Wastes and Race at Twenty 1987–2007," United Church of Christ Justice and Witness Ministries, March 2007, https://www.nrdc.org/sites/default/files/toxic-wastes-and-race-at -twenty-1987-2007.pdf.

191 Covanta waste-to-energy: Goldenberg and Muoio, "Wasted Potential: Recycling Progress in Public Housing Eludes City Officials."

191 is burned: "Facts and Figures About Materials, Waste and Recycling," United States Environmental Protection Agency, https://www.epa.gov/facts-and-figures -about-materials-waste-and-recycling/national-overview-facts-and-figures -materials.

191 minorities or low-income: "U.S. Municipal Solid Waste Incinerators: An Industry in Decline," The New School Tishman Environment and Design Center, May 2019, https://static1.squarespace.com/static/5d14dab43967cc000179f3d2/t/5d5c4bea 0d59ad00012d220e/1566329840732/CR_GaiaReportFinal_05.21.pdf.

192 race is the biggest factor: "Race Is the Biggest Indicator in the US of Whether You Live Near Toxic Waste," Quartz.com, March 22, 2017, https://qz.com/939612/race -is-the-biggest-indicator-in-the-us-of-whether-you-live-near-toxic-waste/.

192 "Stop 'N' Swap" events: "Reduce, Reuse & Recycle at 'Stop 'N' Swap,'" GrowNYC, https://www.grownyc.org/swap.

192 6 percent: Interview with Tiffany Fuller.

192 16.2 percent of textiles: "Advancing Sustainable Materials Management: 2014 Fact Sheet," Table 2, 2.

193 influx of donated goods: Taylor Bryant, "What Really Happens When You Donate Your Clothes—and Why It's Bad," Nylon.com, https://nylon.com/articles /donated-clothes-fast-fashion-impact.

193 **mail-in donation kits:** Jasmine Malik Chua, "With or Without COVID-19, Textile Waste Is on the Rise," *Sourcing Journal*, July 6, 2020, https://sourcingjournal.com/topics/sustainability/textile-waste-coronavirus-u-k-thredup-u-s-fast-fashion-219025/.

193 **in-person donation centers:** Adam Minter, "At Overloaded Thrift Shops, Coronavirus Is Wreaking Havoc," *Bloomberg*, April 29, 2020, https://www.bloomberg.com/opinion/articles/2020-04-30/at-overloaded-thrift-shops-coronavirus-is-wreaking-havoc.

193 **only about 20 percent:** Elizabeth L. Cline, "Tidying Up Has Created a Flood of Clothing Donations No One Wants," May 13, 2019, https://slate.com/technology/2019/05/marie-kondo-tidying-up-donate-unwanted-clothing.html.

193 **"Major" Fred Muhs:** Janice Kiaski, "Salvation Army Has New Local Lieutenant," *Toronto Herald-Star*, July 3, 2017, https://www.heraldstaronline.com/news/local-news/2017/07/salvation-army-has-new-local-lieutenant/.

195 **$17.5 billion in revenue:** "Stats for Stories: National Thrift Store Day: August 17, 2020," U.S. Census Bureau, August 17, 2020, https://www.census.gov/newsroom/stories/thrift-store-day.html.

198 **Garments that are wet:** "Who Is SMART?," Secondary Materials and Recycled Textiles, https://www.smartasn.org/SMARTASN/assets/File/resources/SMART_PressKit Online.pdf.

198 **60 million to 70 million pounds:** Adam Minter, "Global Threads," Resource Recycling, December 2, 2019, https://resource-recycling.com/recycling/2019/12/02/global-threads/.

198 **to places like Panipat:** "Panipat, the Global Centre for Recycling Textiles, Is Fading," *Economist*, September 7, 2017, https://www.economist.com/business/2017/09/07/panipat-the-global-centre-for-recycling-textiles-is-fading.

198 **highest concentration of clothing recyclers:** "We Export . . . ," Bushra International, http://www.bushra-intl.com.

198 **alternative is Pakistan:** "The Story of Asia's Biggest Textile Recycling Hub," DWIJ Products, updated March 6, 2019, https://www.dwijproducts.com/post/the-story-of-asia-s-biggest-textile-recycling-hub.

199 *schmatta* **means "rag":** Jewish-English Lexicon, JEL.Jewish-Languages.org, https://jel.jewish-languages.org/words/477.

199 **somewhere like Star Wipers:** Adam Minter, *Secondhand: Travels in the New Global Garage Sale* (New York: Bloomsbury, 2019), 158.

199 **prices for secondhand wool:** Minter, *Secondhand*, 167–72.

199 **"Take back" programs:** "Recycle Your Denim with Us," Madewell, https://www.madewell.com/inspo-do-well-denim-recycling-landing.html.

199 **Goodwill has their version:** "Reduce, Reuse, Recycle," Goodwill Olympics and Rainer Region, https://goodwillwa.org/donate/sustainability-resources/.

200 **website of I:Collect:** "Partnership Based on Confidence," I:CO, https://www.ico-spirit.com/en/referenzen/partner/.

200 **"innovative, cost-effective":** "Building Textile Circularity," I:CO, https://www.ico-spirit.com/en/company/.

200 **H&M collected 22,761 tons:** "H&M Sustainability," HM.com, 2018, https://about.hm.com/content/dam/hmgroup/groupsite/documents/masterlanguage

/CSR/reports/2018_Sustainability_report/HM_Group_SustainabilityReport
_2018_%20FullReport.pdf.

200 **you'll get 15 percent off:** "Garment Collecting: Be a Fashion Recycler," HM.com, https://www2.hm.com/en_us/women/campaigns/16r-garment-collecting.html.

200 **50 to 60 percent of those donations:** "H&M Sustainability," 50.

200 **H&M was burning:** Jesper Starn, "A Power Plant Is Burning H&M Clothes Instead of Coal," *Bloomberg*, November 23, 2017, https://www.bloomberg.com /news/articles/2017-11-24/burning-h-m-rags-is-new-black-as-swedish-plant -ditches-coal.

CHAPTER 8: PAVED WITH GOOD INTENTIONS: THE END OF THE ROAD FOR CLOTHING IN GHANA

201 **retailer at Kantamanto:** J. Branson Skinner, "Fashioning Waste: Considering the Global and Local Impacts of the Secondhand Clothing Trade in Accra, Ghana, and Charting an Inclusive Path Forward," thesis submitted to University of Cincinnati, March 2019, 11.

201 **sprawls over six acres:** Kenneth Amanor, "Developing a Sustainable Second-Hand Clothing Trade in Ghana," PhD diss., University of Southampton, September 2018, https://eprints.soton.ac.uk/433269/1/LIBRARY_COPY_KENNETH_AMANOR _PHD_THESIS.pdf.

202 **estimated 2,425 tons:** Skinner, "Fashioning Waste," 11, 96.

203 **jeans they got:** "I Went Thrift Shopping at Mantamanto (I Spent What)??," April 29, 2018, https://www.youtube.com/watch?v=_TTs-__ji6Y.

203 **shirts they got:** "Thrifting in Accra Ghana, I Was Shook!!"

203 **There is no shame:** During the pandemic, shoppers' sentiments about secondhand clothes began to reverse. Millennials and Gen-Zers said they were 50 percent more likely to buy secondhand since the pandemic, according to one survey.

203 **Thirty million Ghanaians:** Baden and Barber, "The Impact of the Second-hand Clothing Trade on Developing Countries."

204 **one hundred forty-foot-long containers:** Skinner, "Fashioning Waste," 96–97. Data collected about Kantamanto's inventory comes from the master's thesis by J. Branson Skinner. His data is explicitly not meant to be taken as definitive, given the limits of collection via observation and interviews. That said, it is the most reliable and up-to-date I had access to, given the lack of transparent record-keeping among market players and/or the government.

204 **only seven times:** "Women ditch clothes they've worn just seven times: Items being left on the shelf because buyer feels they've put on weight or they've bought them on a whim," *Daily Mail*, June 9, 2015, https://www.dailymail.co.uk/femail /article-3117645/Women-ditch-clothes-ve-worn-just-seven-times-Items-left-shelf -buyer-feels-ve-weight-ve-bought-whim.html.

206 **A *kayayei* brings each bale:** Interview with Liz Ricketts, The OR Foundation, October 14, 2020.

206 **five per week:** Skinner, "Fashioning Waste," 98.

208 **slashed to make it unsellable:** Skinner, "Fashioning Waste," 105.

208 **"white man is dead"**: Allison Martino, "Stamping History: Stories of Social Change in Ghana's Adinkra Cloth," PhD diss., University of Michigan, 2018, 49.

208 **recorded a denim seller**: "Sorting a Jean Bale in Kantamanto," Theorispresent, December 18, 2018, https://www.youtube.com/watch?v=QFhFBXBP8SU.

209 **make more than $100,000**: Author communication with Liz Ricketts, October 14, 2020.

209 **depend on first selection**: Skinner, "Fashioning Waste," 108.

209 **whereas women sellers**: Skinner, "Fashioning Waste," 110.

211 **Tomato sellers have**: "Kayayo: Ghana's Living Shopping Baskets," video, Aljazeera, https://www.aljazeera.com/program/episode/2018/2/18/kayayo-ghanas-living-shopping-baskets/.

211 **travel a whole kilometer**: Skinner, "Fashioning Waste," 113.

211 **less than $9 US**: One report from the United Nations Population Fund puts this figure higher, finding "Most [*kayayei*] earn an average income of 20–50 cedis per day, approximately 4–10 USD." "Who Are the Kayayei?," UNFPA, 2019, https://ghana.unfpa.org/sites/default/files/pub-pdf/Kayayei%20Photo%20book%202019.pdf.

211 **breaking their necks**: Skinner, "Fashioning Waste," 113.

212 **electronic waste dump**: David Biello, "E-Waste Dump Among Top 10 Most Polluted Sites," *Scientific American*, January 1, 2014, https://www.scientificamerican.com/article/e-waste-dump-among-top-10-most-polluted-sites/.

212 **ten or twelve to a room**: Felix Akoyam, "The Kaya Struggle," October 12, 2016, https://www.youtube.com/watch?v=Rh7aznBetio.

213 **a billion pounds**: Gregory Warner, "The Afterlife of American Clothes," NPR *Planet Money*, December 2, 2013, U.S. International Trade Commission, https://www.npr.org/sections/money/2013/12/10/247362140/the-afterlife-of-american-clothes.

213 **used clothing exports doubled**: Dr. Sheng Lu, "Why Is the Used Clothing Trade Such a Hot-Button Issue?," FASH455 Global Apparel & Textile Trade and Sourcing, November 15, 2018, https://shenglufashion.com/2018/11/15/why-is-the-used-clothing-trade-such-a-hot-button-issue/.

214 **closed off due to travel restrictions**: Minter, "At Overloaded Thrift Shops, Coronavirus Is Wreaking Havoc."

215 **Rwanda's clothing exports**: Abdi Latif Dahir and Yomi Kazeem, "Trump's 'Trade War' Includes Punishing Africans for Refusing Second-Hand American Clothes," Quartz.com, April 5, 2018, https://qz.com/africa/1245015/trump-trade-war-us-suspends-rwanda-agoa-eligibility-over-secondhand-clothes-ban/.

215 **China has captured**: Ruth Adikorley, "The Textile Industry in Ghana: A Look into Tertiary Textile Education and Its Relevance to the Industry," MS diss., Fontbonne University, July 2013, https://www.researchgate.net/publication/274070597_The_Textile_Industry_in_Ghana_A_Look_into_Tertiary_Textile_Education_and_its_Relevance_to_the_Industry.

216 **77 tons of textile**: Skinner, "Fashioning Waste," photo, 11.

218 **turnover has made Kantamanto**: Skinner, "Fashioning Waste," 5.

218 **2014 cholera outbreak**: N. Mireku-Gyimah, P. A. Apanga, and J. K. Awoonor-Williams, "Cyclical Cholera Outbreaks in Ghana: Filth, Not Myth," National

Institutes of Health, https://www.ncbi.nlm.nih.gov/pmc/articles/PMC6003169/. See also *Infectious Diseases of Poverty* 7, no. 51 (June 2018), https://doi.org/10.1186 /s40249-018-0436-1.

218 **Kpone opened in 2013:** L. Salifu, "Draft Final Report. Environmental and Social Audit of Kpone Landfill," Ghana Ministry of Works and Housing, February 2019.

225 **lower greenhouse gas emissions:** "A New Textiles Economy: Redesigning Fashion's Future," Ellen MacArthur Foundation, https://www.ellenmacarthurfounda tion.org/assets/downloads/publications/A-New-Textiles-Economy_Summary -of-Findings_Updated_1-12-17.pdf.

225 **Ten thousand people:** Salman Saeed, Sugam Pokharel, and Matthew Robinson, "Bangladesh Slum Fire Leaves 10,000 People Homeless," CNN, August 19, 2019, https://edition.cnn.com/2019/08/18/asia/dhaka-bangladesh-slum-fire-10000 -homeless-intl/index.html.

225 **more than 2 million acres:** "Brazil's Bolsonaro Says He Will Accept Aid to Fight Amazon Fires," CBS News, August 27, 2019, https://www.cbsnews.com/news/amazon -wildfires-brazil-spurns-20-million-aid-offer-from-g-7-nations-today-2019-08-27/.

226 **46 million acres of land:** Freya Noble, "Government Set to Revise Total Number of Hectares Destroyed During Bushfire Season to 17 Million," 9NEWS (Australia), January 14, 2020, https://www.9news.com.au/national/australian-bushfires-17-million -hectares-burnt-more-than-previously-thought/b8249781-5c86-4167-b191 -b9f628bdd164.

226 **the brink of extinction:** "More Than One Billion Animals Killed in Australian Bushfires," University of Sydney News, January 8, 2020, https://sydney.edu.au /news-opinion/news/2020/01/08/australian-bushfires-more-than-one-billion -animals-impacted.html.

227 **115,000 people who were enslaved:** Rachel Ama Asaa Engmann, "Autoarchaeology at Christiansborg Castle (Ghana): Decolonizing Knowledge, Pedagogy, and Practice," *Journal of Community Archaeology & Heritage* 6, no. 3 (2019): 210, doi: 10.1080 /20518196.2019.1633780.

229 **The RealReal, ThredUp:** Rachel Tashjian, "The Coolest, Most Expensive Clothes on the Planet Are Made from Other, Older Clothes," GQ.com, August 7, 2020, https://www.gq.com/gallery/louis-vuitton-marine-serre-upcycling.

229 **Washington Consensus's aims:** Abdi Latif Dahir, "Used Clothes Ban May Crimp Kenyan Style. It May Also Lift Local Design," *New York Times,* https://www.ny times.com/2020/07/09/world/africa/kenya-secondhand-clothes-ban-coronavirus .html.

CHAPTER 9: LET THE MAKEOVER BEGIN:
TIME FOR A NEW, NEW DEAL

233 *20 percent of their wardrobes:* Ray A. Smith, "A Closet Filled with Regrets," *Wall Street Journal,* April 17, 2013, https://www.wsj.com/articles/SB100014241278873 24240804578415002232186418.

236 **50 percent bigger:** ThredUp, "2020 Resale Report," 2020, https://www.thredup
.com/resale/static/thredup-resaleReport2020-42b42834f03ef2296d83a
44f85a3e2b3.pdf.

238 **presence of a recycling bin:** Jesse R. Catlin and Yitong Wang, "Recycling Gone
Bad: When the Option to Recycle Increases Resource Consumption," *Journal of
Consumer Psychology* (April 11, 2012), doi: 10.1016/j.jcps.2012.04.001.

239 **top twenty fashion companies:** Imran Amed et al., "The State of Fashion 2019: A
Year of Awakening," McKinsey & Company, November 29, 2018, https://www
.mckinsey.com/industries/retail/our-insights/the-state-of-fashion-2019-a-year
-of-awakening.

241 **$30 trillion:** OECD, OECD Business and Finance Outlook 2020: Sustainable
and Resilient Finance, OECD Publishing, Paris, 2020, https://doi.org/10.1787
/eb61fd29-en.

241 **As Sasja Beslik:** Sasja Beslik, "Week 41: ESG Data Is Not Capturing Real-World
Impact," October 11, 2020, https://esgonasunday.substack.com/p/week-41-esg-data
-is-not-capturing.

244 **most lax labor laws:** "Conventions and Recommendation," International Labour
Organization, https://www.ilo.org/global/standards/introduction-to-international
-labour-standards/conventions-and-recommendations/lang—en/index.htm.

244 **border carbon adjustment:** "Reevaluating Global Trade Structures to Address
Climate Change," Council on Foreign Relations, July 2, 2019, https://www.cfr.org
/report/reevaluating-global-trade-governance-structures-address-climate-change.

244 **the case when Rwanda:** "President Trump Determines Trade Preference Program
Eligibility for Rwanda, Tanzania, and Uganda," Office of the United States Trade
Representative, March 29, 2018, https://ustr.gov/about-us/policy-offices/press-office
/fact-sheets/2018/march/title.

246 **"When our measures of development":** Amit Kapoor and Bibek Debroy, "GDP Is
Not a Measure of Human Well-Being," *Harvard Business Review,* October 4, 2019,
https://hbr.org/2019/10/gdp-is-not-a-measure-of-human-well-being.

247 **"disconnection from meaningful work":** Johann Hari, *Lost Connections: Uncovering the Real Causes of Depression* (New York: Bloomsbury, 2018), 76.

247 **"Disempowerment is at the heart":** Hari, *Lost Connections,* 86.

248 **workers on corporate boards:** Andrea Garnero, "What We Do and Don't
Know About Worker Representation on Boards," *Harvard Business Review,* September 6, 2018, https://hbr.org/2018/09/what-we-do-and-dont-know-about-worker
-representation-on-boards.

248 **the Taft-Hartley Act:** "29 U.S. Code §158. Unfair Labor Practices," Cornell
Law School, Legal Information Institute, https://www.law.cornell.edu/uscode/text
/29/158.

249 **America spends half:** Bob Davis, "The Country's R&D Agenda Could Use a
Shake-Up, Scientists Say," *Wall Street Journal,* December 22, 2018, https://www
.wsj.com/articles/the-countrys-r-d-agenda-could-use-a-shake-up-scientists-say
-11545483780?mod=article_inline.

Art Credits

Page 14 "DDT Is Good for Me!" ad: Hunter Oatman-Stanford, *Collectors Weekly*, August 22, 2012, https://www.collectorsweekly.com/articles/the-top-10-most-dangerous-ads/.

Page 30 2017 China vs. US exports: "USA vs China vs India: Everything Compared," WawamuStats, https://www.youtube.com/watch?v=SatG1m0p5g8&feature=youtu.be.

Page 50 World Fiber Production: Bain, "If Your Clothes Aren't Already Made Out of Plastic."

Page 51 Comparative Energy Use in Fiber Production: Business for Social Responsibility, "Apparel Industry Life Cycle Carbon Mapping."

Page 57 The Bangladesh Economy: Fiona Weber-Steinhaus, "The Rise and Rise of Bangladesh—But Is Life Getting Any Better," *Guardian*, October 9, 2019, https://www.theguardian.com/global-development/2019/oct/09/bangladesh-women-clothes-garment-workers-rana-plaza?CMP=Share_iOSApp_Other.

Page 58 Bangladesh Export Growth: "List of Products Exported by Bangladesh," Trade Map—International Trade Statistics, TradeMap.org.

Page 124 The Two Sides of Retail: Justin Fox, "Online Shopping Is Growing, But Isn't Creating Jobs," *Bloomberg*, December 10, 2019, https://www.bloomberg.com/opinion/articles/2019-12-10/retail-jobs-growth-doesn-t-match-expansion-of-online-sales.

Page 182 Donations of Used Clothing in Rich Countries: S. Baden and C. Barber, "The Impact of the Second-hand Clothing Trade on Developing Countries," Oxfam GB (September 1, 2005), https://pdfs.semanticscholar.org/3999/b6470135073bd9c2da9eefe94331212a89aa.pdf.

Grateful acknowledgment is made for permission to reprint the following:

Page 43 The Haze over New York: Neal Boenzi/*The New York Times*/Redux

Page 108 Copyright © GMB Akash/Panos Pictures

Page 210: An image by Natalija Gormalova from the series "Kayayei Sisters." Used with permission.

Page 226: Trading Map of Used Clothes Across the World: *Atlas of the Transatlantic Slave Trade* by David Eltis and David Richardson, map 1 from accompanying web site, "Overview of Slave Trade out of Africa." Published by, and copyright © 2010 by Yale University Press. Used with permission. https://www.slavevoyages.org/static/images/assessment/intro-aps/01.jpg.

Index

Page numbers in italics refer to illustrations.

Abena (Kantamanto retailer), 201, 214, 230

Accord on Fire and Building Safety, 111–15

Accra, Ghana, 179, *180–81*, 201, 217–18, 221, 227–28, 253

Acquah, Sammy, 215, 229–30

Adams, David, 206–10, 214, 230

Adichie, Chimamanda Ngozi, 178

advertising, 151, 158–62, 171, 174–77, 237. *See also* celebrities: fashion endorsements of

AFL-CIO, 88, 140

Africa, xvii–xviii, xx, 161, 199–200, 203, 205, 214, 224, 227, 229

African Growth and Opportunity Act (AGOA), 214–15

air pollution, 41–42, *43*, 44, 55, 84, 191, 225

Akosombo Textiles Limited (ATL), 215

Amazon, xiv, 92, 118
  automated facilities of, 127–35
  blocks unionization, 139–43, 248
  economy, *124*, 131
  executives paid above-market, 138
  fulfillment centers of, 125, 127, 136
  gamification strategies of, 134
  high attrition rate of, 136, 143
  investment in Hollywood, 162–63
  and job creation/job loss, *124*, 131, 146
  as largest US apparel retailer, 119–24
  robots used at, 127–32

Amazon employees, 184, 212
  Black, 141–42
  and childcare, 119, 126
  daily grind faced by, 125–35
  fear warnings/reprisals, 126, 128, 140
  intense physical work of, 126, 130, 134–35, 139
  low pay of, 121–23, 136–39, 155
  offered healthcare plan, 135
  pandemic-related raises for, 123, 130
  pressured to work fast, 126–27, 131, 136
  pushed to become machines, 129, 131–35, 146–47
  UTO allotment of, 119, 126, 128, 132
  at warehouses/fulfillment centers, 121–24

American Eagle, 61, 166

Anderson, Peter, 190–91

Ann Taylor, 61, 120

antiworker laws, 248

apparel broker firms, 94–96, 115. *See also* Li & Fung

Appelbaum, Stuart, 121

Armani, Giorgio, 163

Arnault, Bernard, 122

Arusha, Tanzania, xii

Asia, x, 9, 71, 106, 199

ASOS, 100

Atlantic, The, 142

audits/auditors, 98–103, 105–6, 109–10, 112, 115

Australia, 31, 226

automation, 67, 127–35, 138, 141, 146
Ayivi, Amah, *228*

Bangladesh, 110, 116, 135, 138, 225
  cut-and-sew businesses in, 61–68,
    70–77
  factory collapses in, 92–93, 102–3, 111
  forced/child labor in, 107–8
  improving factory safety in, 111–15
  Korail slum of, *59*, 68–70, 75, 225
  as leading garment exporter, 56,
    58–59, 77
  low-wage manufacturers in, 25,
    53–54
  pollution in, 45, 55, 60–61
  rift between rich and poor in, 58, *59*
  suffers from climate change, 57, 211
  wages in, 71, 105–7
  women protesters in, 72, 82
  women workers in, 81–82, 84, 87, 99
Bank & Vogue (Canada), 214
bankruptcies, 97, 120, 206
Banks, Elizabeth, 165
Baring, Thomas, xx
Barnardos survey, 169
Beauty Vloggers, 166
Beckert, Sven, xvi, 73, 146–47
Becky (Amazon associate), 125, 129–36,
    139–40, 147, 230, 247
Belgium, *182*
Bergh, Chip, 26, 28
Bernays, Edward, 154–63, 170, 174–77,
    231, 250
Beslik, Sasja, 241
Bethell, Steven, 214
Better Buying Index Report, 101–2
Better Work, 106, 110
Bezos, Jeff, 120–22, 130, 137–38, 147,
    155, 163, 165
Black-owned businesses, 234
Black workers, 141–42, 145–46, 150
Blasio, Bill de, 185
Blue River Technology, 18
Boohoo, 167–68
"brand," origin of, 152
Brazil, 79, 99–100

Brooklyn, New York, 183–84, 186,
    190–91, 199, 229
Bruce (Chinese fixer), 37–39
Buffet, Warren, 122
Bureau Veritas (France), 98
business models
  to maximize profits, 27–29, 41, 82, 84,
    102, 138
  merchant brand, 26–29, 87, 97, 123
  shareholder, 27–29, 90, 138, 141,
    241, 248
  stakeholder, 27–28

California, 9, 52, 89–90
Calvin Klein, 85, 95
Cambodia, 45, 53–54, 74–75
Canada, 87, 90, *182*, 198, 204, 214, 223
cancer, 3, 17, 40, 43, 46–48, 144, 175
capitalism, xv–xvii, xix–xx, 24, 116,
    144–47, 159, 243
carbon
  and burning of trash, 191, 225
  decreasing use of, 239–40
  and farming, 18, 20
  limiting emissions of, 243–44
  produced by clothing industry, 35–36,
    49, 51, 244
  sequestered from atmosphere, 20,
    243, 251
Carothers, Wallace H., 49
Carson, Rachel, 15
Case, Anne, 138–39
caste system, 142
CDC, 172
celebrities, 175
  fashion endorsements of, 162–67,
    173, 237
  powerful impact of, 229, 232, 239
  red-carpet industrial complex and,
    163–64, 169
  special note to, 238–39
  stylists, 151, 163–65
CEOs, 113, 143, 145, 240, 242
  in age of hyperglobalization, 94–95
  high compensation of, xvi, 28, 76,
    137–38, 248

reducing compensation of, 117
and wages of average workers, 76,
137, 246
See also Bezos, Jeff
Cerf, Vint, 162
Chanel, 163, 196
chemical use, 83, 251
in clothing manufacturing, 47–48
in cotton fabric production, 34, 36, 45,
48–49
decline in, 239–40
exposure to, 38, 47–48, 53, 87
in farming, 3–5, 13–19, 21
pollution from, 41–42, *42*, 44–45
regulations/standards for, 48, 53, 249
child
labor, 81, 96, 100, 107, *108*
slavery, 78–79
childcare, 70, 114, 119, 137, 140, 143,
246–47, 249
China, x, 1, 61, 64, 84, 95, 116, 204, 244
cotton production in, xiv, 22–23, 36
dominates clothing industry, 53, 77,
144, 247
dominates manufacturing, 24, 29–30
energy emissions of, 35–36
fabric quality control, *33*, 34
forced labor in, 8, 36, 79, 107–8
garment exports of, 36, 56, 92, 213
modernization of, 53, 249
moves away from clothing industry,
53, 56
online advertising of, 167
online shopping in, 170
open-door policy of, 24–25, 29, 31, 37,
41, 215
pollution in, 39–42, *42–43*, 44–45
47, 53, 87, 221
recreates Ghana textiles, 215
Six Priorities of, 29, 248
textile industry of, 29–32, 50
unions illegal in, 109
China Water Risk, 39–40
cigarette campaigns, 174–76, 237
circularity, in fashion, 228, 236–39
Citizens of Humanity, 89

citizenship, xiii, xxi, 231–51
City of New York Department of
Sanitation (DSNY), 183–89
Claire (Paris shopper), 148–51, 155,
171, 176
Clean Clothes Campaign (CCC), 71, 96,
106, 110–12
Clean Water Act (1972), 43
climate change, 41, 51, 241
and clothing in landfills, 191, 217, 225
and clothing manufacturing, 35–36,
53–54, 57, 245
and farming, 19–21, 211
policies for, 243–44
and shopping online, 170–71
Cline, Elizabeth, 232
Clinton, Bill, 89
clothes shopping
addiction to, 237
becoming better at, 11, 233–34
constant exposure to, 171
healthy relationship with, 239
and marketing manipulation, 151–54
reexamining habits of, 149–51, 230
as security, 168–70
as tool for social control, 155–57
and unconscious desires, 154–55,
157, 252
See also online shopping; retail: therapy
clothing
gross overproduction of, 213
positive relationship with, 232–33
post-WWII demand for, 13
rewearing of, 165, 225, 229–30, 236,
239, 251
clothing brands
bad purchasing practices of, 102–5
connecting farmers to, 20
demand transparency from, 20, 115, 235
develop low-wage manufacturers, 25
and manufacturing partners, 95–105
and power over factories/suppliers,
103–4, 113–14
and reducing their impact, 235–36, 244
and supplier relationship, 103–4
clothing fibers, trade in, 25

coal-powered manufacturing, 53
codes of conduct, 62, 95–102, 104–6,
    108–9, 113, 251
Colbert, Jean-Baptiste, 153
Colombo, Sri Lanka, 73–74, 213
colonialism, xvi–xvii, 152
Common Thread: *Vogue* x Amazon
    Fashion, 121
Conscious Closet, The (Cline), 232
consumerism, 128, 178, 250
consumers, 95, 116, 151
    can address wrongs of system, 231–41
    and citizenship, 250
    and engineered consent, 158–59, 161,
        174–76
    and farm production methods, 20
    questioning risk of synthetic fabrics, 53
    and "the power of your purse," 232–39
    and "the power of your voice," 232,
        239–42
    on truth of clothing brands'
        practices, 115
    and unconscious connections/desires,
        154–57, 243
consumption
    and buying used clothing, 236–37
    and celebrity advertisements, 166–67
    cycles/patterns of, 151, 176
    harmful to our health, 170–73
    postwar, 156–57
    reexamining habits of, xx–xxi, 214
    slowing down of, 246–47
    social-modeling based, 160, 162, 167
Cooke, Rob, 73
Coons, Chris, 79
Corporate Duty of Vigilance law (2017),
    79–80
corporation social responsibility (CSR),
    102, 107–10, 115, 214, 250
corporations, 242
    environmental impact of, 241–42
    outsourcing jobs by, 138–39
    put workers on boards, 91, 113, 248
    rising economic/political power of,
        138–39, 143
    See also CEOs

Corruption Perceptions Index, 92
cosmetics, 6, 9, 166
cotton
    clothing of, xvii, 6, 11, 28, 31,
        50, 80
    dyeing of, xiv, 22, 31, 34, 44–45, 80
    fabric, xviii, 31, 34–36, 44–45, 48–49,
        61, 215
    harvesting of, *8*, 9–10, 12, 80
    leading producers of, 57
    plant/boll of, xviii, *7–8*, 12, 32
    post-WWII demand for, 13
    price of, 7, 10
    shipped overseas, 28–29, 31
    spinning of, 22–24, 29–31, 80, 161
    textiles of, 22, 80
    trade in, xvii–xviii, xx, 25, *226*
    weaving of, xiv, xix, 9, 22–23, 31–32,
        33, 34, 80, 251
    See also farming: of cotton
cotton production
    at cotton gins, xviii–xx, *9*, 10, 22,
        31–32, 80
    drove industrialization/inequality,
        xix–xx
    energy use in, 51
    fueled slave labor, xvii–xx, 65, 81–82,
        242, 244
    global, xiv
    hazardous chemicals used in, 15–17
    lack of transparency in, 10, 19
    and pesticide-resistant seeds, 15–16
    in United States, 1–4, 15, 26, 31, 50
    and women's labor, xvi
    worldwide, 1–2, 4, 15, *50*
cottonseed oil, 9
Council of Fashion Designers of America
    (CFDA), 120
Covanata waste-to-energy facility, 191
COVID-19 pandemic, xx, 130, 132, 137,
    140, 144
    and bankruptcies, 97, 120
    clothing donations increase in, 193
    and decline in apparel imports, 36
    and decline in clothing sales, 247
    and decline in donated goods, 214

and Ghana's *kayayei*, 212
and increase in social media use, 165
and increased consumption,
    246–47
online sales increase in, 120–23, 170
and organic farmers, 11
and our relationship to things,
    149–51, 162
retail stores close in, 103–4
trash removal and, 185–86
crude oil, xi, 1, 50–51. *See also* fossil
    fuels
cut-and-sew business, 80
    and average workers' wages, 71
    in Bangladesh, 61–66, 70–77
    in El Paso, 89
    outsourcing of, 56, 246
    process of, 61–64
    women workers in, 63–65, 69–77,
        81–82, 87

Dandeniya, Ashila, 101, 117
Danu (Sri Lankan garment worker),
    74–75, 107, 135, 213, 230
"Dark Side of the American Dream"
    (Kasser), 159–60
Deaton, Angus, 138–39
Debroy, Bibek, 246
Del Rey, Jason, 121, 123, 140–41
democracy, xv, xx–xxi, 145, 154,
    156–57, 243
denim, x, 9, 15
    Bangladesh top exporter of, 56
    full-package production of, 89
    in landfills, *221*, 222
    leading producers of, 77
    made in El Paso, 26–28, 86
    origins of, xvii
    overseas production of, xiv, 26–28, 37,
        56, 84
    reducing price of, 28–29
department stores, 61, 81, 121
Depop (UK), 236
Dhaka, Bangladesh, 55–60, 68–73,
    92–93, 128, 132, 221, 222
Dickson, James, 50

Dior, 163, 178
discarded clothing, 179–83, 232–33,
    242, 244, 252–53
distribution, xv, xix, 121–22, 131, 177,
    242, 245–46
donated clothing, 252
    dumped/burned, 173–74, 180–81,
        216–17, 224
    exported, *182*, 204–5, 213–15, 224
    what happens to it, 192–200
    See also landfills; Salvation Army
drought, 6–7, 11, 17, 211
DuPont, 49–50, 52
dyes/dyeing, xi, xiv, 31, 34, 44–45,
    80, 152

East India Company, xvii–xviii
Easterling, Addison Rae, 166
Economic Policy Institute, 123
economic power, 29, 138–39, 143
economies
    fashion industry key to, 152–54
    growth-focused, 32, 41, 56–57, 75–76,
        84, 249
    industrial, 144
    opening-up of, 24–25, 31
ecosystems, 17, 19, 45, 52
Edelman Trust Barometer, xx
Edison, New Jersey, 124–29
efficiency engineers, 128
Efficiency Movement, 63
effluents, 96, 221
    from cotton fabric production, 34,
        44–45
    from dyes, 44–45
    government monitoring of, 87
    harm caused by, 34, 46, 48
    impact on China's rivers, 39–40,
        44–45
El Paso, Texas, 26–28, 30, 84–90
energy industries, 43
energy use
    of cotton fabric production, 35–36,
        44–45
    of fashion industry, 35–36,
        169–70, 244

environmental
  codes of conduct, 96, 99–100,
    104–5
  compliance (costs of), 110
  concerns of agriculture, 19–20
  consciousness, 4, 161, 169, 249
  degradation, xiii, 147, 173, 192
  offenses, 77–80
  organizations, 250
  protections/regulations, 43–44, 92,
    103, 113–16
environmental impact
  of clothing in trash, 191, 217–19,
    224–25
  of clothing industry, xv–xvi
  of clothing production, xi–xiii
  of corporations, 241–42
  of donated clothes, 180
  of fashion industry, 231, 238–40, 251
  of producing organic cotton, 12–21
  of purchases/brands, 150–51,
    169–70, 175
Environmental Protection Agency (EPA),
    15, 43–44, 92, 192
Erin Brockovich (film), 40–41, 43
ESG, 241–42
Ethical Trading Initiative (ETI), 98–99
ethics, 73–74, 97–99
Ethiopia, 53–54, 56, 58
Europe, 41, 154, 244
  consumer patterns in, 116, 167
  impact of globalization in, 145
  puts cap on importing textiles/
    garments, 25
  secondhand garments of, 204
  and slave/cotton trade, xvii, xix,
    226, 227
  unions in, 90–91, 140
European Chemical Agency, 38
European Union, 35, 213
Everlane, 97–98, 105
Ewen, Stuart, 157

Facebook, 172
Fairport, New York, 189
Farah, Inc. (El Paso), 88

farmers, xiv, 10–11, 20–21, 116,
    245, 251
Farmers National Company, 21
farming, xi, xiv, 20, 50, 116, 243
  affected by climate change, 211
  and Black workers, 141
  conventional, 5–10, 13–19
  of cotton, xiv, xvii, 1, 4–6, 15, 19, 21,
    36, 51, 115, 222
  with effluent-tainted water, 39, 46, 47,
    84, 106
  industrial, 3, 13, 15
  post-WWII, 13, 49
  subsidies for, 7, 11
  See also organic: farming
Fashion Nova, 167
fashion season/calendar, 153–54, 167,
    177, 239
Fashion Week, 111, 120, 176
fast fashion, 49, 98, 149, 151, 167–69,
    196, 202, 235, 242
feminism, 178
fertilizers, 3–5, 12–19, 48–49, 158
fiber production, 50–51
finance/financial institutions, xviii,
    241–42
financial insecurity, xx, 121, 137, 159–61,
    168, 242, 245–46
flood, 17, 116, 211
Forbes, 245
Fortune 500, 189
fossil fuels, 35, 49–53, 161, 191, 243
FRAME, 89
France, xvii, 215, 248
  auditing firms in, 98
  fashion industry of, 152–54, 176, 247
  government-regulated sales of, 149
  greenhouse gases in, xv, 35
  human rights/environmental abuses
    laws of, 79–80
  returned goods destroyed in, 174
Frank, Robert, 165
freshwater, 45–46
Freud, Sigmund, 154–55, 157, 162
Friedman, Milton, 27
Frontlash, The, 232

Gap, 61, 92, 94, 97–98, 100, 105,
     123, 238
garment factories, 92, 96–104
   and brands' purchasing price practices,
        109–10
   cut wages (Bangladesh), 106–7
   improving safety of, 111–15
   noncompliance with wage laws, 105–7
   price/turnaround pressures on,
        106–7, 115
   unsafe conditions at, 110–11
garment workers
   as human machines, xix, 61–69,
        135, 245
   pressured to work fast, 72–73, 103, 170
   See also cut-and-sew business; specific
        countries
Gates, Bill, 76, 122
Gazipur District (Bangladesh), 60–68, 96
GDP, 57, 59, 90, 137–38, 157, 245–46
gender
   bias, 62–65, 164–65
   wage gap, 246
German Catalogue of Textile
        Auxiliaries, 45
Germany, 3, 13, 91, 144, 249
   auditing firms in, 98
   Covariant robot of, 131
   and environmental impact of
        clothing, 150
   greenhouse gases in, xv, 35
   have workers on corporate boards, 248
   returned goods destroyed in, 174
   and used clothing, 182, 213
   in WWII, 154–55
Ghaffary, Shirin, 121, 123, 140–41
Ghana
   cholera outbreak in, 218
   connected to Western world, 226–27
   electronic waste dump in, 212, 221
   pollution in, 212, 217, 225
   secondhand garments exported to,
        xiv–xv, 213–14
   suffers from climate change, 211
   textile/clothing manufacturing in,
        215, 229

See also Accra, Ghana; Kantamanto
     market (Ghana); Kpone (Ghana)
     landfill
Gifford, Kathie Lee, 96
gig economy, 107
global
   clothing production, 57, 104,
        109, 213
   cotton production, xiv
   economy, xv, xviii, 92, 121, 141, 253
   fiber production, 50
   labor organizations, 111–12
   population increase, 246
   trade relations, xiii
   See also trade: global
globalization, 44, 77–78, 84, 94
   job losses due to, 89–90, 246
   promises of, 115–16
   working-class impacted by, 145
GMO seeds, 15–16
Goodwill, 195, 197, 199, 207
Google, 45, 143
Great Affordability Crisis, 137
Great Britain, xvii, xix–xx, 204, 227
green energy, 144
greenhouse gases, xv, 35–36, 191,
        225, 241
Greenpeace, 44
Greer, Linda, 44
Guangdong, 36–39, 42, 44–46, 85
Gucci, 163, 228

H&M, 92, 167, 196, 202
   and clothes in landfills, 179, 223
   committed to circular fashion,
        237–38
   paid for orders in pandemic, 104
   "take back" programs of, 199–200
   and workers' rights/wages, 76,
        111–12, 121
Ha-Meem factory fire, 111
Hamiton Avenue Marine Transfer
        Station, 186–88
Haney, Rick, 19–20
Hari, Johann, 134–35, 247
Harvard Business Review, 246

health risks, 46, 79, 247
  from dyes, 34, 44
  faced by Amazon employees, 126,
    130–35
  for garment workers, 71, 171
  and illegal dumping of textiles, 218
  from incinerator emissions, 191–92
  of *kayayei* (girl porters), 211
  for retail workers, 171
  for sanitation workers, 185–86
  and shopping online, 170–73, 175
  See also air pollution; water pollution
healthcare, 246–47, 249
Hearst, William Randolph, 162
herbicides, 5, 14–16, 18–19
High Price of Materialism,
  The (Kasser), 159
Hilton, Paris, 166
Hong Kong, 25, 31, 53
housing, 119, 125, 137, 141, 246, 249
human brains, 29, 67, 134, 158–60, 171,
  173, 175–76
human rights offenses, 77–80
Human Rights Watch, 103
hyperglobalization, 27, 90–94, 112,
  114, 242

immigrants, 81–82, 145, 196, 199
India, 25, 66, 106
  cotton trade in, xvii, xix–xx, 1, 6,
    19, 222
  exports textiles/fabrics, 19, 24, 61
  secondhand clothing market in, 236
  secondhand garments graded in,
    198–99, 204
Inditex, 76, 97–98, 100, 104
Indonesia, 96, 106
industrial
  accidents, 82–83, 92–93, 111–12
  economies, 144, 152
  engineering, 63–67, 187
  policy, 144, 146, 153, 246–49, 252
Industrial Revolution, xviii, 80–81,
  108, 227
industrialization, xvi, xviii–xx, 41,
  157–58, 215

IndustriALL, 111–12
inequality
  and cotton production, xviii–xx
  in fashion industry, xvi, 130
  in income, 58, 242, 245–46
  racial, xvi, 249
influencers, 151, 176
  impact of, 232, 239
  on Instagram, 149, 229, 237
  and materialist tendencies, 160, 173
  role of, 166–69, 229
  special note to, 238–39
insecticides, 5, 14–16
Instagram, 115, 153, 169
  fashion ads on, 94, 181, 237
  influencers on, 149, 229, 237
  and materialist tendencies, 160
  used by Ghanaians, 204
International Finance Corporation
  (IFC), 110
International Labour Organization
  (ILO), 103, 106, 110, 113, 116
International Monetary Fund, 215
Isenberg, Nancy, 146

Jacob Javits Center (New York), ix–x,
  xiii, 202
Japan, 25, 95, 144, 208, 233, 249
J.Crew, 92, 97–98, 105, 120, 123
jeans
  chemicals in, 49
  cost to make them, 28
  major role in global economy, xv
  manufactured overseas, 28–30,
    37–40, 84
  stone-washed look of, 85
  and wage of average workers, 251–52
  wash and finish of, 37–40, 84–89
  "whiskering" of, *86*
Jessica (trend junkie), 168–69
Jewish immigrants, 81–82, 199
Jim Crow laws, 141
job loss, 88–90, 131, 140, 146, 246
John Deere company, 18
journalists, 96
"Juki Girls," 74

Kahl, Kent, 15–16, 22–23, 31
Kahn, Robert E., 162
Kantamanto market (Ghana)
    and custom design shops, 227–30
    dumps textile waste, *216–19*, 222
    kayayei (girl porters) of, 202, *210*,
        211–13, 221
    retailers in, 215–17
    sells secondhand clothes, 201–4, *205*,
        206–9
    socioeconomic value of, 224
    women sellers of, 201, 209, 214
Kapoor, Amit, 246
Kasser, Tim, 159–62, 168
Kenya, 214, 229
Kiabza (India), 236
King, Martin Luther, Jr., 142
Kondo, Marie, 193, 232
KonMari six-rule method, 233
Kpone (Ghana) landfill, 179, *180–81*,
        204, *216*, 217–19, *220–21*, 222,
        223, 224–25, 230, 232

labor, xx, 89, 142
    day labor, 74–75, 107
    and economic/political power,
        138–39
    forced, 36, 79, 107–8, 113
    international standards for, 77–80,
        91–93, 116
    laws, 77–80, 87, 244
    movement of, 80–83, 86, 144
    offshored, 90, 142, 146
    outsourced, 138–39, 198
    See also child labor; codes of conduct;
        slave labor
Labor Standards Act posters, 87
landfills, xiv
    chemicals in, 219–20
    clothing in, 174, 179–82, 189–93,
        199–200, 216–25, *221*, 222, 225
    fires in, *220*, 221–25, 232
    laws concerning, 192
    in United States, 186, 190–92, 218
    See also Kpone (Ghana) landfill
Latino workers, 150

Laura (Amazon associate), 118–22,
        125–40, 143, 147, 230, 247–48
leather, xi, 49, 61
legislation, 143–44, 146
    against cigarette ads, 175
    for minimum wage, 89–90
    for sustainable farming practices, 11
    for workers' rights/protections, 82–83,
        141, 248
Lesotho, 106
Levi's, 26–28, 30, 85, 88, 91, 94, 123,
        200, 208, 228
Li & Fung, 94–96, 115
life expectancy decline, 138–39, 246
lobbying, 7, 11, 25, 87, 90, 102, 113,
        142–43, 250
Lokko House (Ghana), 227, *228*, 229–30
London, England, 222, 253
Lost Connections (Hari), 247
Louis XIV (Sun King), 152–54, 156, 167,
        215, 248
Louisiana Purchase, xx
Lowell, Massachusetts, 81–82
Lubbock, Texas, 1, 9–10, 31
Lululemon, 99
LVMH Family, 122

Macy's, 102, 121
Mad Men, 151, 158, 175, 177, 178
Madewell, 97–98, 199–200
Malaysia, 79
Manfreda, Stefania, 227
marketing, 26–27, 84, 94, 159, 167
    about sustainability, xi–xii, 235
    and circularity, 236–39
    customize ads, 177–78
    manipulation of, 151, 176, 241
    and our consumption habits, xv, 151,
        214, 234
    and "take back" programs, 199, 214
Marks & Spencer, 104, 112
Martineau, Philip, 28
materialism, 159–63, 168, 171–73
Mazur, Paul, 158
McKinsey & Company, 31, 150, 170

McKinsey Reports, 167
mechanized work, 73, 84, 130–32. *See
also* automation
media, xvi, 89, 96, 104, 109, 111, 137,
175, 252
Mellis, Keith, 186–89
mental health issues, 160–61, 170–73,
234, 246–47
merchant brand model. *See* business
models
Met Gala, 120, 165
methane gas, 191, 225
#MeToo, 164
Mexico, 9, 28, 87, 214
microplastics, 51, 225, 233
middle class, 90, 143, 158, 213, 249
mills, 22–24, 81–82. *See also* textile:
factories/mills
mind-numbing work, 118, 177, 246–47
minorities, xvi, 145, 191–92
Monsanto, 16
Morgan, Anne, 82
Muhs, Fred, 193–96, 200, 208, 230
Müller, Paul, 14
Multifiber Arrangement (MFA),
25–29, 95

National Soil Health Institute, 17
Native American land, xx, 242, 244
natural fibers, 13, 45, 51–52, 233
natural resources, 151, 170, 181
Natural Resources Defense Council
(NRDC), 44
neoliberalism, 106, 111
criticism of, 92, 245–46
dismantling of, 112, 114, 116
extreme capitalism in, 144–47,
159, 243
and shareholder primacy, 27–29, 138
tenets of, 90, 97, 102
US policies of, 215
Netherlands, xvii, *182*
New Deal, 83, 108, 141, 144–45, 248–49
New Jersey, 189, 190–91
New Standard Institute (NSI), xiii–xiv,
39, 67, 80, 165, 240–41

New York City, 150, 202, 222, 253
clothing donations in, 192–96, *197*
garbage collection in, 183–92, 217
rag dealers in, 198–99
sweatshops in, 96
women protesters in, 81–83
New York Times, The, 99
NGOs, 110–12, *182*, 201, 208
Nike, 96, 100–101, 179, 222
North American Free Trade Agreement
(NAFTA), 87, 89
North Carolina, 23

oceans, plastics in, 51–52
online shopping, 119–23, 149–52, 159,
170–75, 196. *See also* Amazon
OR Foundation, 201–2, 208
organic
certification, 4–5, 12, 48
cotton, 4–21, 19, 48
farming, 5–8, 12, 16–19, 243
standard, 5, 48, 242–43
Organic Trade Association survey, 12
Organization for Economic Cooperation
and Development (OECD), 91
Ortega, Amancio, 76
OSHA, 92
Osu Castle (Ghana), 226–27
overtime, 71–72, 96, 103, 106, 108, 123,
136, 141

Pakistan, 25, 61, 198
Paris, France, 77, 148–49, 153–54, 176
particulate pollution, 191, 225
Pattinson, Robert, 163
Pennsylvania, xi, 189
Pepper, Carl, 1, *2*, 3–23, 31, 115, 230
Pepper, Kelly, 4
Pepper, Leslie, 3–4, 15, 17
Pepper, Terry, 3–4, 17
Perkins, Frances, 83
pesticides, 3–5, 12, *14*, 15–19, 48–49,
158, 220
PET (polyethylene terephthalate), 50
Pew Research Center, 172
plastics, 49–53, 225. *See also* microplastics

political
    candidates, 89, 145–46, 247
    engagement, 155–58, 160, 250, 252
    power, 138–39, 143
Politico, 146
polyester, xi, 32, 49–51, 61
polyvinyl acetate (PVA), 45
potassium permanganate, 38
poverty, xiii, 53, 57, 116, 145–46, 191
pricing trends, 102–5
profits
    maximizing of, 27–28, 138
    the only goal, 27–29, 67, 73, 102
    of top fashion companies, 239–40
    of a very few, 245
Progressive Era, 108
protectionism, 153, 215, 244
protests, worker-led, 72, 82–83,
    92–93
psychology, modern, 154–55
PVH, 104, 111

Qing Mao Weaving, Dyeing and Printing
    Co. (China), 22–24, 30–32, *33*,
    34–40, 44, 51, 61, 80
Qingshi, Xu "Dashi," 40–45, 48, 51, 54,
    60, 230

race-wage gap, 244–46
racism, 145–47
    structural, 142, 192, 234, 239, 242,
        244–45, 252
    and union workers, 141–42
rags, 198–200
Rahman, Shahidur, 66
Rana Plaza (Bangladesh factory), 92–93,
    102–3, 111
ready-to-wear garments, 81–82
Readymade Sustainability Council, 112
Reagan, Ronald, 159, 175
recycling, 192, 195, 198–200,
    225, 238
Red Cross, 199
regenerative agriculture, 18–21, 251
renewable energy, 247, 251
Rent the Runway (RTR), 168–70

Republicans, 89
retail
    centers, 202, 222
    during pandemic, 120–23, 149–51
    and shift to digital stores, 119–24, 128,
        149–51, 170
    therapy, 155–56, 174, 251–52
Retail, Wholesale, and Department Store
    Union (RWDSU), 121
returned goods, 174
Ricketts, Liz, 201–3, 208–9, 212,
    221, 224
Rima (garment worker), 69–77, 84, 106,
    135, 163, 225, 230
Roberts, Julia, 163
Robinson, Joan, 115–16
robots, 127–32, 146
Rodrik, Dani, 92, 112, 114, 145,
    247, 249
Roosevelt, Franklin D., 83
Roundup, 15–16
Rwanda, xii, 214–15, 244

Sabel, Charles, 145
Saco-Lowell, 22
Salvation Army, 180, 193–96, *197*, 198,
    202, 208, 214, 238
Sam (Amazon associate), 126, 128–31,
    134–36, 140, 147, 247
Sawyer, Diane, 111
Saxbe, Darby, 172–73
Schmidt, Eric, 143
Schneiderman, Rose, 82
screentime, 170–72
Secondary Materials and Recycled
    Textiles Association, 197–98
secondhand garments, 192, 245
    banning imports of, 214–15
    dumped in Ghana landfill, 216–25
    export of, xiv–xv, 197–98, 213–14
    grading of, 197–98, 204–5, 208
    improving system for, 224–25
    reexamining market of, 213–16
    reinvention of, 227–30, 253
    renewed interest in, 150–51, 193
    resale of, 201–19, 236–37

secondhand garments (*cont.*)
  shipping bales of, 204, *205*, 208–9
  slaves of the system, *210*, 211–13
  sorting of, 206–8
  See also donated clothing
severance packages, 88–89, 91
sex trafficking/sex industry, 74–75
Shanghai, 36
Shaoxing, China, 22, 39, 40, *43*, 44–45, *47*, 222
"shoddy blankets," 199
Shuler, Liz, 140
Silent Spring (Carson), 15
silk, 49, 57, 95, 176, 208
Silk Road, 25
single-use garments, 167–70, 238
Skinner, J. Branson, 201–2, 209, 217
slave labor, xx, 73, 145, 152
  and brand/vendor codes of conduct, 96, 98
  foundation of fashion industry, 141, 242
  fueled by cotton industry, xvii–xx, 8, 65, 244
  and making of cotton garments, 81–82
  modern-day, 78–79, 107–8
  roots of, xv–xvi
  trade in, *226*, 227
smartphones, 92, 125, 133, 149, 161, 165, 170–72, 207, 209
social media, 149, 151, 155, 162, 165–69, 237, 240, 243, 252
social responsibility, 27, 97–98, 102, 105, 171, 199, 214, 250
soil health, xiv, 3, 5, 12–14, 17–20, 243, 251
South Carolina, xi, 189
Spain, 152–53
Sri Lanka, 92, 104–5, 116, 221
  garment workers of, 73–75, 81, 84, 135, 213
  wages in, 71, 105–7
  woman demands worker's rights in, 101, 117
Star Wipers (Ohio), 199
starch, 32, 34, 46

stocks/stock market, 90, 122, 137–38, 175, 248
style, your own, 169–70, 176–78
stylists. *See* celebrities: stylists
suffragist cause, 82
suicide rates, 6, 138, 172, 246
supply chains, xiv–xv, 76, 95–96, 98, 105, 241, 247
sustainable
  brands of clothing, 36, 97–98, 104, 150
  farming, 11–12, 18, 21, 243
  fashion, xi–xii, xiii, 232, 235–36, 241
  See also regenerative agriculture
Sustainable Development Goals, xii–xiii
sweatshops, 96, 101, 107
synthetic fabrics/fibers, 45, 49–53, 191, 199, 225. *See also* polyester

Taft-Hartley Act, 248
"take back" programs, 199–200, 214, 238
Tamara (hairstylist), 155–58, 172–74, 177, 190, 234
Target, 100, 104
Tashjian, Rachel, 176
Taylor, Frederick Winslow, 63
technological advances, 7–8, 13, 18–19, 80–83
Tema, Ghana, 204–5, *218*
Texas, xiv, 1–2, 4, 8, 22, 89–90, 145, 220, 222. *See also* El Paso, Texas
Texas Organic Cotton Marketing Cooperative, 4, 10–11
textile
  factories/mills, 24, 30–32, *33*, 34–46, 48, 81, 251
  industry, 25, 46, 152–53, 215
  production, 25, 35–36, 81–82
  workers, 25, 47–48, 53, 56, 58, 87, 251
Thailand, 25, 95
Thatcher, Margaret, 159
The RealReal, 229, 236
Thomson Reuters Foundation, 79
ThredUp, 193, 196, 229, 236
TikTok, 153, 160, 166–67
T.J.Maxx, 156, 158, 173–74

Tommy Hilfiger, 95, 111
toxic
  chemicals, 38–40
  gases, 45, 225
  waste, 190, 212, 219–21, 252
trade, 27, 36, 44, 54, 114, 244
  agreements, 84, 87, 89–92, 109,
    113–15, 144
  global, xiii, 24–25, 41, 84, 144, 152,
    242, 249
  networks of, xvii–xviii
  open-door, 29
  restrictions, 87, 215
  of slave laborers, *226*, 227
  of used clothing, *226*, 227
Transparency International, 92
trash, 184–92. *See also* landfills
Triangle Shirtwaist Factory, 82–83, 93
Trump administration, 41, 90, 214–15
Trump, Donald, 89
Turkey, 6, 61

Underhill, Paco, 177
Underwriters Laboratories, 98
unemployment, 142, 144, 150
unions, 79, 184, 242, 252
  advantages/successes of, 140–42,
    144, 146
  and Black workers, 141–42
  and clothing brands, 111–13
  and collective bargaining, 88, 90, 109,
    141, 143
  demands of, 87–88
  ensuring representation in, 244–45
  and garment factories, 108–10
  impeding formation of, 138–43, 248
  and textile/apparel workers, 25, 82–83
  in US versus Europe, 90–91, 140
United Kingdom, xv, 35, 63, 150,
    168–69, 194, 213, 215, 236
United Nations (UN), xii–xiii, 45, 78
United States
  clothing manufacturing in, x–xi,
    26–28, 86
  and donations of used clothing, *182*
  exports used clothing, 213
  needs to invest in industry/labor,
    144–45
  overproduction of clothing in, 213
  prohibits slave-made imports, 78–79
  rich-poor divide in, 90, 122, 137–38,
    145, 165, 245–46
  shifts to a "desires culture," 158–59
  shifts towards retail economy, 123
Universal Declaration of Human
    Rights, 77
US Committee on Public
    Information, 154
US Congress, 79, 89, 175
US Customs and Border Protection, 79
US Department of Agriculture (USDA),
    2, 5, 19–20, 119
Used Clothing Exports, 198

Vestiaire Collective (France), 236
Veterans Association, 173
VF Corporation, 104
Vietnam, 53–54, 56, 61, 92, 106, 107–8
Viramontes, Cesar, 84–91, 109, 230
Virginia, 186, 189–91
Vito (sanitation worker), 183–86,
    188–89, 230
Vogue, 121
voting, 145–46, 161, 250

wages, 102, 110
  in foreign countries, 25, 58, 100,
    106–7
  gap in, 137–38, 142, 164–65, 244–46
  for garment workers, xv, 74–75, 89,
    105, 115
  legal protection for, 86–87
  "living," 71, 76, 106, 137, 145
  local industry standards for, 105–6
  minimum, 89–90, 105–8, 141, 244–45
  unions bargain for, 88, 90, 140–41
  in United States, 25, 90
  universal laws on, 77–78
Wai, Jacob, 31
Walmart, 95–96, 100, 102, 119, 196
Wang, Charles, 30–32, 37, 39
Wang, Yvonne, 31

warehouse workers. *See* Amazon employees
wash houses, 26, 38–39, 84–90
Washington Consensus, 215, 229
Waste Management, 189–90
waste regulation, 40–48
water pollution, 218, 225, 251, 253
  in Bangladesh, 60–61
  in China, 39–41, *42*, 43–45, *47*,
    60–61, 87
  from clothing/textile industries,
    44–46, 48
  from effluents, 44–46, 48, 87
  and food/farming, 84, 106
  from plastic fibers, 51–53
  in United States, 43, 192
water scarcity, 45
Waverly, Virginia, 190–91
Wayss, Rob, 111–13
We Should All Be Feminists (Adichie), 178
wealth, xv–xvi, 76, 84, 137–38, 144–47,
    152, 159, 165
weapons factories, 13–15, 157–58
West Africa, 201, 203, 224
Whinfield, John, 50
white supremacy, 146
*White Trash: The 400-Year Untold History
    of Class in America* (Isenberg), 146
Whitney, Eli, xviii
Why We Buy (Underhill), 177
Wilkerson, Isabel, 142
Wilson, Woodrow, 154, 157
Wintour, Anna, 121
women garment workers, xv–xvi, xviii
  assault of, 72–74, 82, 87, 99
  in cut-and-sew business, 63–65, 69–77,
    81, 87
  exploitation of, 213, 221
  lead labor movement, 80–83
  protests of, 72, 82–83, 86
  work barefoot, 65, 87
  See also Kantamanto market (Ghana):
    kayayei; sex trafficking/sex
    industry; *specific countries*

wool, xi, xvii, xix, 50–51, 199
worker protections, 34, 62, 65
  and codes of conduct, 96, 99–100,
    104–6
  and COVID-19, 168
  high-level enforcement of, 111–13
  lack of, 25, 87, 89, 93, 100,
    130, 139
  legal protection for, 86–87
  policies/laws for, 91–93, 109, 245
  and trade agreements, 84, 89, 109
  in United States, 91–92, 139
Worker Rights Consortium, 111–12
workers' rights, 77–91, 96, 101, 108–9,
    113, 116, 141–42, 244. *See also*
    unions
working hours, 82–83, 86–87,
    96, 102
workplace safety, 77–83, 86–87,
    96, 108, 110–15, 251
World Bank, 215
World Health Organization
    (WHO), 173
World Trade Organization (WTO), 29,
    31, 215
World War I, 13, 154
World War II, 3, 13–14, 25, 49, 154–55,
    157–58
Wrangler, 85, 91, 228

Xinjiang, China, 8
Xintang, China, 30, 37

YouTube, 166, 203

Zady, ix–xiv, 96, 202, 235
Zara, 76, 97, 99–100, 167, 171, 196, 203,
    228, 241
Zhaolu Environmental Protection Center
    (China), 40–42
Zhejiang province, 22, 51
Zimbabwe, 79
Zuckerberg, Mark, 122